The Colorado Front Range

The Colorado Front Range
A Century of Ecological Change

Thomas T. Veblen
Diane C. Lorenz

University of Utah Press
Salt Lake City Utah

∞ The paper in this book meets the standards for permanence and durability established by the Committee on Production Guidelines for Book Longevity of the Council on Library Resources

Library of Congress Cataloging-in-Publication Data

Veblen, Thomas T., 1947–
The Colorado front range : a century of ecological change / Thomas T. Veblen, Diane C. Lorenz.
p. cm.
Includes bibliographical references.
ISBN 0-87480-351-9 (pbk. : alk. paper)
1. Botany—Colorado—Ecology—History—Pictorial works. 2. Botany—Front Range (Colo. and Wyo.)—Ecology—History—Pictorial works. 3. Vegetation dynamics—Colorado—History—Pictorial works. 4. Vegetation dynamics—Front Range (Colo. and Wyo.)—History—Pictorial works. 5. Front Range (Colo. and Wyo.)—History—Pictorial works. I. Lorenz, Diane C., 1957– . II. Title.
QK150.V38 1990
581.5′2642′097886—dc20 90-31547
CIP

Contents

Figures

Acknowledgments

The research on which this book is based was funded by the Council for Research and Creative Works of the University of Colorado, the Colorado Commission for Higher Education, and the American Association for State and Local History. We thank all of these institutions for their support. For field assistance we thank K. Cirillo and N. R. Johnson. For permitting us to use some of their photographs we are indebted to H. E. Malde and N. R. Johnson. For granting permission to reproduce photographs from their collections we thank the Denver Public Library, the U.S. Geological Survey Photographic Library, the University of Colorado's Norlin library, the Rocky Mountain National Park Historical Collection, the Boulder Historical Society Collection of the Carnegie Branch Library for Local History, the Colorado State Historical Society, and F. Crossen. We also thank the staff of CCL Photo Lab of Boulder and Larry Harwood of the University of Colorado for their expert printing of many of our photographs.

For critically commenting on the manuscript we are grateful to P. Klein and K. S. Hadley.

Introduction

The comparison of present vegetation to its condition at the time of white settlement in the western United States relates to many modern problems and controversies in conservation and vegetation management. An important example of a contemporary vegetation-management problem is the controversy over fire suppression versus "let burn" policies in our national parks and national forests. Another example is the debate over past and present impacts of livestock, and whether livestock should be permitted in wild lands managed by the federal or state governments. Planning for appropriate use and management of renewable natural resources requires predictions of how these resources might change under current use patterns. In the absence of long-term experiments, which require many decades of monitoring before useful results are obtained, historical information on vegetation response to past use and management is essential. Effective vegetation management must be based on a thorough understanding of how past human activities such as logging, burning, and livestock raising have contributed to the present vegetation patterns.

Natural disturbances, such as lethal insect outbreaks that may occur at rather infrequent intervals, also play major roles in creating landscape patterns. Knowledge of the roles of these natural disturbances is essential for effective resource management. To attain long-term goals of managing or preserving vegetation in wild landscapes, managers must understand the complex interactions of human-caused disturbances, changed patterns of vegetation, and altered patterns of natural disturbances. Similarly, the general public must understand these complex interactions if they are to make intelligent choices among policy alternatives. These considerations apply not only to wilderness and park areas where the principal management goal is preservation but also to forest reserves where long-term stability under multiple use for timber, watershed protection, wildlife protection, and recreation is the management goal (Lotan et al. 1985).

The prevalent contemporary perception of national parks and wilderness areas in the western United States is that, in addition to providing recreation, such reserves must preserve the vegetation in a state similar to what it was at the time of white settlement in the late nineteenth century. In these reserves, natural change (including changes triggered by the Native American population before Euro-American settlement), may be acceptable but changes caused by Euro-American civilization are not (Chase 1986; Vale 1987). Thus, particularly in wilderness and parks, it is essential that we have a realistic perception of what the vegetation was at the time of white settlement. Similarly, in the extensive wild lands fringing urban areas in the western United States the public needs to know how past use and management policies have influenced the present landscape. It is only with such a historical perspective that the public can make informed choices among the variety of vegetation-management options proposed by government agencies and special-interest groups.

The Colorado Front Range, the easternmost range of the Rocky Mountains, is an ideal area for a historical study of vegetation change. The Front Range contains large areas where preservation of pristine vegetation conditions is the primary goal. These include Rocky Mountain National Park in the northern part of the range and the Indian Peaks Wilderness in the north-central part of the range. It also has a large area in the Roosevelt and Arapaho national forests where multiple use is the accepted criterion of vegetation management. Here, a balance among the interests of timber production, wildlife protection, watershed protection, and recreation is sought. Significantly, the Front Range contains a growing population along the urban corridor from Colorado Springs to Fort Collins, which is placing increasing and often conflicting demands on the forest resource. These demands are not confined to the national forests and the national parks but also include private, state, county, and city properties where the condition of the vegetation is a major public concern. Desire for "open space" and recreational demand has created a whole new series of vegetation-management issues over the past two decades. These are areas where homeowners have a self-interest in fire suppression and control of deer and other wild-animal populations. However, attaining these management goals is likely to have a long-term impact on the character of the open-space vegetation. These wild-land urban fringes are where policies of fire management and wildlife management are likely to conflict most frequently with other land uses.

Our goal in this book is to document, through the use of repeat photography, some of the major changes that have occurred in the Front Range over approximately the past 100 years. Repeat photography is the practice of finding the location of an earlier photograph, reoccupying the original camera position, and taking a new photograph of the same scene (Rogers et al. 1984). The technique of repeat photography is a useful way of showing major changes in a landscape. We hope the information provided here on vegetation changes in the Front Range will be of use to the citizen who wants to be better informed for resource-management issues. We also hope that this photographic history will be a source of enjoyment to people who simply are interested in the past and present vegetation patterns of the Front Range.

Vegetation Dynamics

The focus of this book is on change in the structure and composition of vegetation, in particular of the forest cover. Change in vegetation is universal. On short time-scales, there are changes from week to week associated with the seasons. Thus, flowering, elongation of twigs, growth of new leaves, increase in tree height and diameter, are all examples of predictable seasonal changes in vegetation described as phenological changes. The concern here is with change in plant populations resulting from the basic demographic processes of birth and death. Any patch of vegetation is a dynamic phenomenon due to these demographic processes. At an annual time-scale there is the fluctuation in population sizes of herbaceous plants that complete their life cycle during a single year. At the much longer time-scale of centuries to millennia, change becomes evident that may be linked to climate change, long-distance migration of plant species, soil development, or evolutionary change in the genetic characteristics of a species. Both these time-scales, annual variation and millennial changes, are beyond the scope of this photo-historical study. Instead, our concern is with the changes of an intermediate time-scale that are detectable over periods of a few decades to a century or two.

We are concerned with changes in vegetation structure and composition that are evident in the comparison of photographic records over time spans of about 50 to 100 years. Where the change has resulted in a different species composition it is termed "plant succession." Plant succession is usually defined as a directional, cumulative change in the relative species abundances that occupy a particular area (Miles 1979; Barbour et al. 1987). For long-lived woody plants the time required for detection of such change is at least a decade or two. Some of the change described in the photo-historical comparison of Front Range sites is properly described as plant succession.

Much of the change that has occurred over the past 100 years in the Front Range is not successional change but simply change in forest structure without major shifts in species composition. Forest structure is defined by the ages, sizes, and spatial arrangement of the tree populations of a forest stand. For example, where a burned stand has regenerated to the same species, the structure is likely to have shifted towards a stand of relatively small, young trees growing in substantially greater density than in the original stand. Or, the result of a small blowdown may be a patch of young stems conspicuously denser than the surrounding large trees of the same species. Thus, structural changes do not necessarily result in a cumulative change in the species composition of a stand. However, they are still visually obvious changes that give a forested landscape much of its character. We use the term "vegetation dynamics" to include both successional change and structural change. Vegetation dynamics includes both successional changes in composition and structural changes not resulting in a major shift in species composition. Where it is clear that there has been a cumulative change in the species composition of a stand, such change will be termed plant succession. Often, however, in the absence of detailed historical descriptions of the vegetation of the site it is difficult to detect such successional changes.

Our preference for the term "vegetation dynamics" also reflects the fact that most modern students of vegetation change reject much of the theory and terminology that developed around the concept of plant succession early in this century. Traditional plant successional theory in North America was developed largely by the botanist Frederick E. Clements. Clements (1916) viewed succession as a highly orderly and predictable process in which successional change represented the life history of a climax plant community. According to this viewpoint, the climax was a condition of great stability in which the vegetation had reached an equilibrium with the present climate. Disturbance such as fire, in this view, was relegated to the role of a large-scale triggering event followed by a long disturbance-free period of constant climate during which a series of predictable plant communities replaced one another on the same site. Thousands of studies of successional change have demonstrated that succession is not as orderly and predictable as believed by Clements (White 1979). Although some degree of prediction is possible in most habitats, random events such as seed dispersal make prediction difficult. Traditional successional theory is also seen as not viable because in many forested landscapes the time period required for succession from a pioneering community to a climax community may be so long that the assumption of a constant climate is not reasonable. Over the many centuries required for a climax forest of long-lived conifers to develop, climatic variability may affect the course of vegetation change. Similarly, the emphasis on a long disturbance-free period is not consistent with what we now know about vegetation dynamics (Pickett and White 1985). Thus, in many landscapes successional change may not proceed all the way to the climax end point simply due to the frequency of disturbance and the slowness of compositional change involving long-lived plants. The more modern view of vegetation dynamics may be termed a "dynamic" or "kinetic" view in which there is no assumption of either long-term site stability or of the existence of an end point to succession (Drury and Nisbet 1971; Veblen et al. 1980). Thus, the modern view of vegetation change emphasizes the importance of repeated, relatively frequent disturbance and accepts continuous change in vegetation as the norm (Pickett and White 1985).

Despite the past overemphasis on the climax concept in traditional successional theory, some plant communities have population structures indicating that the dominant species have long been regenerating at that site. To avoid the theoretical implications associated with the term "climax," we prefer to describe such communities as in "compositional equilibrium" or "steady-state" (Whittaker 1975; Veblen in press). For stands in compositional equilibrium or steady-state, there should be young, intermediate, and old trees of each species. Many other stands have tree-age population structures indicating that they are not in compositional equilibrium. These are stands in which some tree species are represented only by old individuals while other species occur as young individuals. Such stands appear to be undergoing successional change. In Clements' terminology (1916), such stands are referred to as "seral" to indicate that they are part of a "sere" or sequence of successionally related plant communities.

Vegetation change results from environmental changes induced by the plants themselves and from external environmental changes. Examples of the former include the changes in soil properties or understory light conditions that may create microsite conditions more favorable to some plant species and less favorable to others. Examples of the latter include environmental changes such as climatic fluctuations and those that are commonly regarded as "disturbances," such as fire, droughts, landslides, snow avalanches, insect outbreaks, and many other natural or human-caused alterations of the environment. Plant ecologists commonly refer to the plant- or community-induced changes as "autogenic" and the environmental changes that originate externally to the plant community as "allogenic" (Tansley 1935; Barbour et al. 1987). Autogenic change refers to the modification of microsite conditions, which in turn may result in a site of increased or decreased suitability for certain species. These autogenic changes tend to be gradual changes and are relatively predictable. Autogenic change also includes the modifications of microsites associated with the senescence and death of individual trees, which in turn may result in structural and/or compositional change in the stand. Allogenic changes may be either sudden or gradual. The most important kind of gradual allogenic change is slow climatic change, which usually is difficult to detect over time spans shorter than many decades. More sudden allogenic changes, or disturbances, are much less predictable than autogenic changes but in the Front Range have played a major role in shaping the present landscape.

Disturbance may be defined as "any relatively discrete event in time that disrupts ecosystem, community, or population structure and changes resources, substrate availability, or the physical environment" (Pickett and White 1985). Disturbances are rather sudden events that damage or kill plants, thereby releasing some of the resources otherwise used by those plants. This release of resources may result in enhanced growth of nearby competing plants or may create the opportunity for establishment of new individuals. Disturbances that originate principally outside a plant community are termed exogenous disturbances and may be important allogenic factors of vegetation change. Common examples of exogenous disturbances resulting in allogenic vegetation change in Front Range forests include fire (both natural and human-set), logging, and lethal insect outbreaks. In contrast, endogenous disturbances are events that result primarily from the growth and death of the plants themselves. Thus, the most common type of endogenous disturbance in forests is the fall of an individual, senescent tree. In reality, occurrence and the characteristics (e.g., intensity, area affected, etc.) of a particular disturbance event are the consequence of the interaction of the properties of a plant community with its external environment. For example, the probability and intensity of a fire are influenced both by fuel availability, which results primarily from the age of a stand, and by weather conditions that affect the moisture status of that fuel and provide an ignition source (lightning). Similarly, lethal insect attacks depend both on stand characteristics (i.e., tree vigor and tree size) and on external factors such as weather conditions. Despite the difficulty of classifying some disturbance events as either endogenous or exogenous, the dichotomy between disturbances that are mainly controlled by external forces versus those controlled by the community itself is still a useful one.

It is important to recognize that vegetation change is usually the result of both autogenic and allogenic factors and of endogenous and exogenous disturbances. To understand the patterns and processes of vegetation change in the Front Range, we need to identify the autogenic and allogenic factors that are important to vegetation change in this landscape. We also need to understand the responses of the important tree species to their environments, and we need to know the history and characteristics of the disturbances that have affected this landscape.

The Study Area

Physical Geography

The Colorado Front Range is the easternmost range of the Rocky Mountains, extending from near Cañon City northward to the Wyoming border (Chronic and Chronic 1972). This is the longest continuous range in the state, and it sharply demarcates the western limit of the High Plains (Fig. 1). On the eastern margin, steeply dipping sedimentary rocks mark the limit between the Rocky Mountains and the High Plains at an elevation of about 1700 m. The western margin, although not as sharply defined, consists of a precipitous decline into the broad troughs of North and Middle Parks (Griffiths and Rubright 1983). The northern end of the Front Range merges with the Medicine Bow Mountains and the Never Summer and Mummy Ranges, which extend towards the northwest (Ives 1983). At its southern end, the Front Range plunges into the plains, although a southwest-trending ridge connects it with the Wet Mountains south of Cañon City (Chronic and Chronic 1972). The focus of this study is on the northern two-thirds of the Front Range, comprising the mountainous portions of Boulder and Larimer Counties.

Along the highest portion of the Front Range, the Paleozoic and Mesozoic sediments have been eroded away to expose a complex series of Precambrian crystalline rocks (Lovering and Goddard 1950). Thus, the mountain core of gneiss, schist, and granite, which contains many economic mineral deposits, is exposed throughout much of the Front Range. Several large masses of granite resulting from the slow cooling of intrusions of molten rock at great depth make up portions of the core of the Front Range. Examples include the Boulder Creek Granite and the Silver Plume granite. Along the eastern flanks of the range, the eroded edges of the sedimentary rocks that once covered the entire range are exposed. Examples include the steeply inclined sedimentary formations such as Denver's Red Rocks Park and the Flatirons near Boulder (Chronic and Chronic 1972). Tilted layers of Paleozoic and Mesozoic sediments form the cuestas and hogback ridges that are so conspicuous at the juncture of the High Plains with the Front Range.

The crest of the Front Range reaches 3800 to 4200 m in elevation, rising some 2500 m above the High Plains in an east-west distance of only 30 km. This crestline forms the Continental Divide. The eastern slope of the Front Range consists of long, steep slopes leading up to interfluves that range from sharp ridges to broad, nearly level upland surfaces. These belts of benchlike surfaces, as well as the rolling uplands of the terrain traversed by Trail Ridge Road in Rocky Mountain National Park and the almost flat tops to mountains such as Longs Peak and Mt. Evans, are thought to be remnants of ancient erosion surfaces (Chronic and Chronic 1972; Ives 1983).

Much of the higher part of the Front Range extending from the crestline down to about 2600 m elevation was glaciated during the Pleistocene (Richmond 1960; Benedict 1973). The work of the glaciers is seen in the cirques,

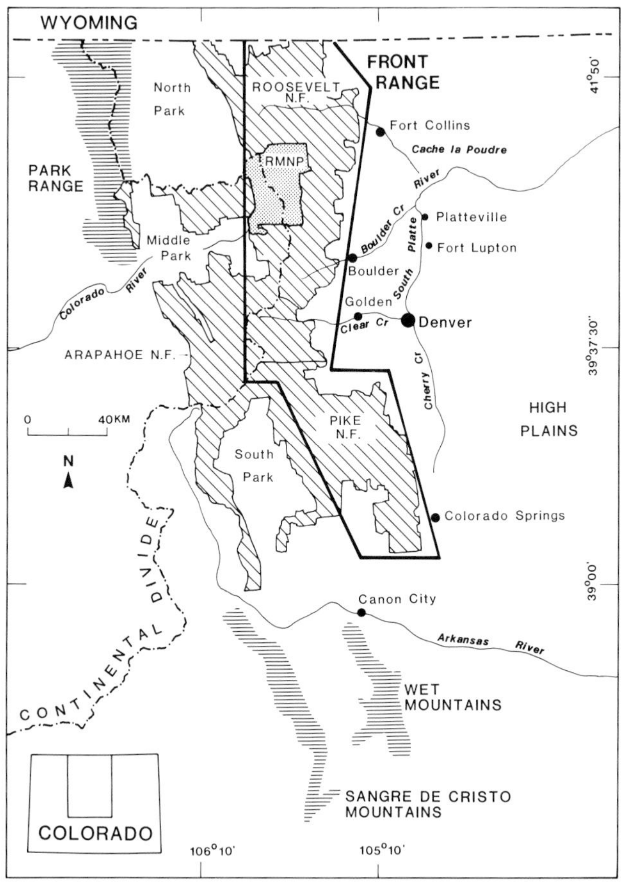

Figure 1. Map of the Colorado Front Range (delineated by the thick line). Abbreviations: RMNP, Rocky Mountain National Park; NF, National Forest.

tarns, the deep U-shaped troughs of the upper courses of river valleys, and broad moraine-enclosed basins. Indeed, a major reason for establishing Rocky Mountain National Park in 1915 was the preservation of the scenic grandeur of its glacial setting.

The soil pattern in any mountainous area is expected to be complex due to the likelihood of substantial changes over short distances in soil-forming factors—parent material, climate, topography, vegetation, and time. Soils of the Front Range are mostly poorly developed, slightly acid, coarse-textured, and usually very rocky (Peet 1981). Steep topography and rapid natural erosion tend to limit the time available for soil development, so soils on many sites are shallow. At high elevations, the development of soils from glacial till deposited by the most recent Pleistocene glaciation has also restricted the time available for soil development. The soils of the Front Range have been described by Johnson and Cline (1965) and Marr (1961) according to elevational zone. Further information on soils is provided in the description of the vegetation of the Front Range.

The climate of the Front Range is controlled by a number of factors. Its interior location results in relatively dry conditions and wide differences between summer and winter temperatures. Its position relative to cyclonic storm tracks moving on the average from west to east accounts for most of the winter precipitation. The rugged relief of the Front Range interacts with air masses to produce complex climatic conditions. Three types of air masses have a strong influence on the climate of the Front Range (Griffiths and Rubright 1983). Two are sources of moist maritime air and one is a source of cold, dry arctic air. Northern Pacific maritime air masses periodically bring cool moist air to the region during the non-summer months. Tropical maritime air masses that form over the Gulf of Mexico bring warm humid air to the Front Range primarily during summer. Cold, high-pressure air masses originating over northern Canada and the Arctic Ocean occasionally bring dry and extremely cold air masses to the Front Range during winter. As a major barrier to the free flow of air masses, the Front Range inhibits or modifies the movement of air masses by blocking or deflecting the air upward, downward, and sideward (Hansen et al. 1978). In winter, cold polar air spreading southward over the High Plains, for example, often is arrested against the eastern slope, resulting in colder temperatures on the adjacent plains than at higher elevations in the Front Range. Pacific air masses entering the region from the west are forced upward by the Front Range, often causing heavy snowfalls in the mountains while to the leeward side of the range clear weather may prevail. Similarly, the northwestward flow of humid Gulf air is often arrested by the Front Range, triggering heavy showers on the eastern slope.

The cyclonic storms that flow across Colorado most of the year are greatly modified by the mountainous topography. Where stalled cyclonic storms create upslope movement of moist air masses along the eastern slope of the Front Range, high precipitation results. Precipitation from October through May is primarily associated with cyclonic storms and at higher elevations occurs mainly in the form of snow. During the summer months, cyclonic storms are less frequent and precipitation occurs as intense thunderstorms often accompanied by hail and lightning. Maximum precipitation usually occurs in May, with secondary peaks in July and August reflecting the influence of summer convective storms (Barry 1972).

The climate of the Front Range is continental in character, with wide differences in temperature between summer and winter. On a daily basis, the weather is subject to sudden and sometimes extreme changes. Along the elevational gradient from the High Plains towards the crestline of the Front Range, marked changes in average temperatures and precipitation occur (Barry 1972). Mean annual precipitation increases slowly at first as one ascends the foothills and then more abruptly towards higher elevations. Mean wind speeds also increase dramatically with elevation, from 2 to 4 m per second at the foot of the eastern slope to over 10 m per second near the crest (Barry 1972). During winter, when a high-pressure cell stagnates over the Great Basin, strong, dry Chinook winds may attain wind speeds in excess of 100 mph at the foot of the eastern slope. At elevations over 3000 m monthly maximum wind speeds exceed 100 mph for most of the year (Glidden 1982).

The Vegetation

Marr (1961) classified the vegetation of the eastern slope of the Front Range into a series of "climax regions" based primarily on elevation (Fig. 2). Peet (1981) described in detail the variation in the forests along environmental gradients of elevation and moisture in Rocky Mountain National Park. In the following description, Marr's (1961, 1964) elevation-based scheme is adopted because of its simplicity and widespread use. Many of the details of the variation in the vegetation in relation to environmental gradients, however, are taken from Peet (1981).

Marr (1961) divided the vegetation of the eastern slope of the Front Range into five elevational zones, which he characterized as "climax regions" (Fig. 2). Although many ecologists today object to some of the implications of the climax concept, such as the emphasis on stability as opposed to continued change, these climax regions are a useful way of summarizing the major patterns of vegetation in the Front Range. Marr's five elevational zones are as follows: plains grassland (below 1710 m), lower montane (1830–2350 m), upper montane (2440–2740 m), subalpine (2840–3350 m), and alpine tundra (above 3480 m). The upper and lower montane zones are often combined to form a single montane zone. Marr recognized substantial zones of transition, or ecotones, between each pair of elevational zones.

Most of the plains grassland region has been drastically modified by cultivation, heavy livestock grazing, and urban and suburban development. There remain only rare patches of relatively undisturbed grassland on the plains adjacent to the Front Range. The important native grass species include both big and little bluestem (*Andropogon gerardi* and *Schizachyrium soparium*), needle grass (*Stipa comata*), and side-oats grama (*Bouteloua curtipendula*).* Little bluestem stands, needle grass, and grama grass along with linear-leaf wormwood (*Artemisia glauca*) are most common where the soil is rocky and shallow. Thickets of wild plums (*Prunus* spp.) and hawthorn (*Craetagus* spp.) grow in ravines as do cottonwoods (*Populus* spp.) and willows (*Salix* spp.) along stream courses. Closer to the mountains are patches of shrubs such as mountain mahogany (*Cercocarpus montanus*), skunkbrush (*Rhus trilobata*), buckbrush (*Ceanothus fendleri*), and bitterbrush (*Purshia tridentata*).

The vegetation of the lower montane zone consists of open forest of broad-crowned conifers frequently interrupted by grasslands. Ponderosa pine (*Pinus ponderosa* var. *scopulorum*) and Douglas fir (*Pseudostuga menziesii*) are the dominant trees of the lower montane zone. There is a striking contrast in stand density on north- as opposed to south-facing slopes (Fig. 3). The mesic, north-facing slopes support much denser stands. Ridgetops and gently rolling uplands are usually well forested with ponderosa pine and less

*The botanical nomenclature in this chapter follows Weber (1976). For species with widely known common names, scientific names are given only at the first mention.

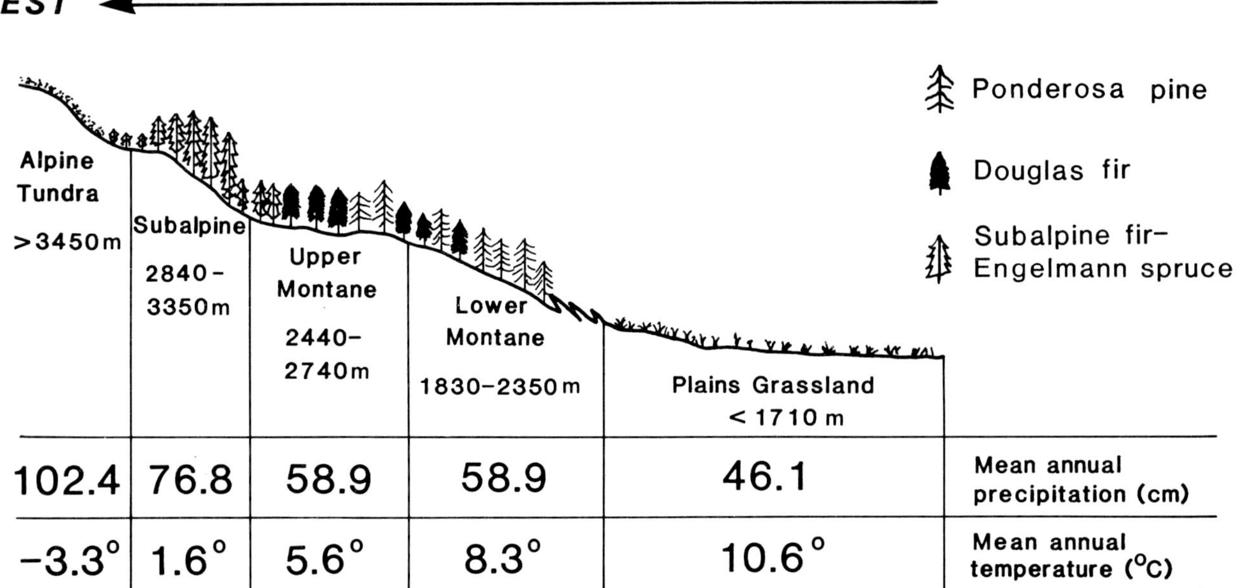

Figure 2. Schematic diagram of vegetation zones and climatic parameters on the eastern slope of the Front Range. Data from Barry (1972) and Marr (1961).

commonly with Douglas fir. On south-facing slopes stands are much more open and the proportion of Douglas fir much less.

The most common forest type of the dry sites of the lower montane zone is open woodland of ponderosa pine, characterized by scattered trees covering less than 50 percent of the ground. The open understories are dominated by grasses. They grade into plains and narrow-leaf cottonwood (*Populus sargentii* and *P. angustifolia,* respectively) forests along stream courses and into stands of Douglas fir or mixed ponderosa pine and Douglas fir on mesic, north-facing slopes. The lower margin of the ponderosa pine woodlands grades into grassland characterized by more fine-textured soils (Peet 1981). Along the lower limits of its distribution, ponderosa pine is conspicuously associated with rocky and coarse-textured soils while fine-textured soils are dominated by grasses. Scattered, small ponderosa pine are typical of many of the grasslands and suggest a recent invasion by trees. Shrubs commonly associated with ponderosa pine in the lower montane zone include mountain mahogany, skunkbrush, and bitterbrush. The small Rocky Mountain juniper (*Juniperus scopulorum*) is commonly present, particularly on the drier sites, but is rarely of significant size or abundance.

The distinguishing features of Marr's upper montane zone include the gradual increase in importance of Douglas fir and the presence of aspen (*Populus tremuloides*), limber pine (*Pinus flexilis*), and lodgepole pine (*Pinus contorta* var. *latifolia*) (Marr 1961). On mesic, north-facing slopes at 1700–2200 m and on xeric, south-facing slopes and ridgetops at 2400–2800 m mixed stands of ponderosa pine and Douglas fir are common (Peet 1981). They form a transition from the open woodlands of lower elevations to the dense, closed forests of the higher elevations. In comparison with the ponderosa pine woodlands, stand density in forests characterized by Douglas fir is much greater. On the more mesic sites of the upper montane zone, pure, dense stands of Douglas fir occur. In the upper montane zone, limber pine occurs mostly in mixed stands with the other conifers on sites that are particularly inhospitable due to shallow, rocky soils and/or exposure to strong winds. Aspen grows both on xeric ridgetops, where it occurs as mainly dwarfed individuals, and on mesic sites, where dense stands of tall trees occur. Both limber and aspen are much more characteristic

Figure 3. A typical contrast of dense Douglas fir forest on a north-facing slope (left) and an open stand of ponderosa pine (right) in Four Mile Canyon.

of the subalpine zone than of the upper montane zone. Similarly, lodgepole pine rarely occurs at elevations below 2600 m (Marr 1961). Colorado blue spruce (*Picea pungens*) forms streamside forests principally in the upper montane zone. We will typically use the term "montane" to refer to elevations where ponderosa pine and/or Douglas fir are the most common trees, thus combining Marr's lower and upper montane zones.

The principal forest types of the subalpine zone include lodgepole pine–dominated stands towards lower elevations, and Engelmann spruce (*Picea engelmannii*) and subalpine fir (*Abies lasiocarpa*)–dominated stands towards higher elevations. Lodgepole pine forests occur mainly between 2400 and 3200 m, occupying sites of intermediate moisture conditions. Typically, these are relatively young, dense stands with dark, dry understory conditions (Fig. 4). The paucity of understory plants in these forests is striking. Usually only sparse herbs and scattered shrubs such as common juniper (*Juniperus communis*), huckleberry (*Vaccinium myrtillus*), and kinnikinnik (*Arctostaphylos uva-ursi*) can be found in these stands. Towards the lower end of its distribution lodgepole pine occurs in association mainly with Douglas fir. Towards the upper end of its distribution lodgepole pine occurs in association mainly with Engelmann spruce and subalpine fir. On the drier sites, subalpine fir is the principal associated species, while on moister sites Engelmann spruce is also common. Lodgepole pine also occurs in association with aspen, particularly in young postfire stands.

Most of the subalpine zone is characterized by Engelmann spruce and subalpine fir–dominated forests (Fig. 5). Small numbers of lodgepole or limber pine are often also present, particularly on the drier sites. Spruce–

Figure 4. A typically sparse understory beneath a dense postfire stand of lodgepole pine at an elevation of about 2900 m.

and subalpine fir–dominated stands occur on all but the most xeric sites above 3100 m, and in cool, sheltered valleys at elevations as low as 2500 m. The relative dominance of the two canopy tree species and the understory composition vary substantially over a gradient from excessively moist to xeric sites (Peet 1981). Open bog forests occur on the limited areas of level, poorly drained terrain above 2800 m. These sites are characterized by waterlogged soils and thick accumulations of organic material. These bogs are a mosaic of forest patches and wet meadow dominated by sedges (*Carex* spp.). In Peet's (1981) classification, the wet spruce-fir forest type is differentiated from bog forests by its occurrence on mineral substrate rather than accumulated organic debris. This type also occurs on nearly flat terrain near timberline. Engelmann spruce often attains its greatest size with diameters of 1 m or more in this type. Given the slightly better drainage of the wet spruce-fir type, the distribution of trees is more uniform and the understory is characterized more by forbs than by sedges. The forbs typical of this type include *Senecio triangularis, Mertensia ciliata, Saxifraga odontoloma, Trollius laxa,* and *Epilobium* spp., and the common shrubs are subalpine prickly currant (*Ribes montigenum*) and huckleberries (*Vaccinium* spp.).

The mesic Engelmann spruce–subalpine fir type occurs on cool, sheltered, but well-drained sites above 2700 m and is one of the most widespread forest types of the subalpine zone. The most important understory species are the semishrubs (i.e., low-growing shrubs that die back to the ground in winter) myrtle blueberry (*Vaccinium myrtillus*) and broom huckleberry (*Vaccinium scoparium*), and the forbs *Arnica cordifolia, Pyrola secunda,* and *Epilobium angustifolium*. Open slopes above

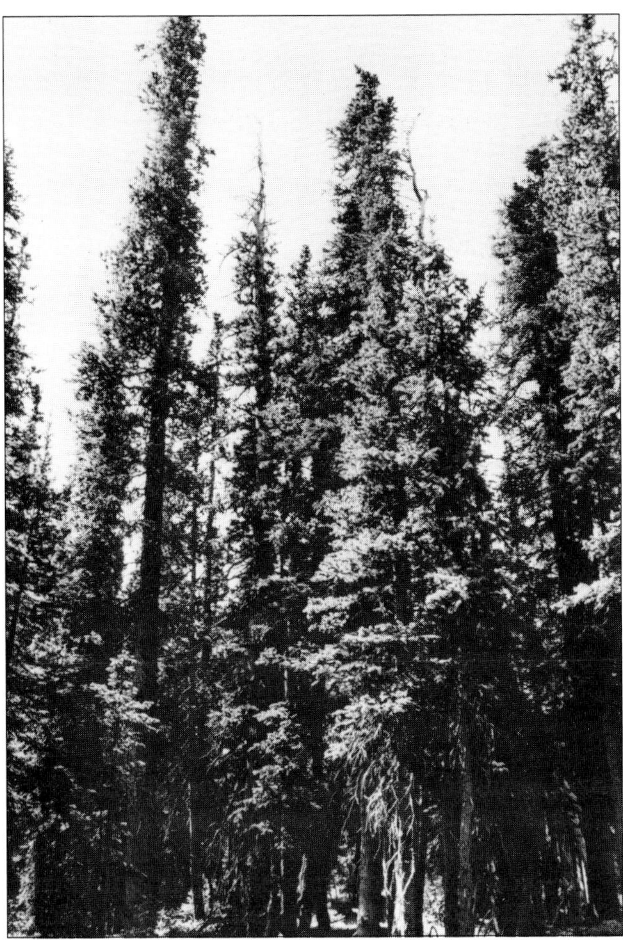

Figure 5. A stand of Engelmann spruce and subalpine fir at an elevation of approximately 3100 m.

Figure 6. A limber pine growing on a morainal ridge at an elevation of about 3000 m. Note the large lower branches, which indicate that this tree originally grew under open conditions. In this stand Engelmann spruce and subalpine fir are currently replacing limber pine and creating a much denser stand.

3100 m are typically characterized by Peet's (1981) xeric spruce-fir type. Maximum tree sizes are somewhat smaller in this type than in the mesic type, with spruce rarely being larger than 50 cm diameter in the former and often exceeding 70 cm diameter in the latter. The understory is primarily dominated by *Vaccinium* spp., but in this type, broom huckleberry is much more common than myrtle blueberry, which is more common in the mesic spruce-fir type. Occasional, scattered individuals of lodgepole and limber pine also occur in this type, particularly on the more xeric and lower elevation sites. The xeric spruce-fir type is the most extensive forest type of the subalpine zone.

Towards lower elevations, the Engelmann spruce–subalpine fir types give way, often along abrupt fire-induced boundaries, to lodgepole pine–dominated forests. Towards higher elevations, the spruce-fir types gradually open up into alpine meadow and alpine tundra. In the transition between forest and alpine tundra, islands or clusters of stunted, poorly formed krummholz ("twisted wood" in German) of spruce and fir are common. In the krummholz zone, tree form ranges from dense, low mats to severely flagged but upright individuals. Here, subtle interactions of snow persistence and wind have a dominating influence on the physiognomy of the vegetation.

Limber pine dominates stands on xeric sites above 2800 m, particularly sites that have shallow, rocky soils and are exposed to strong westerly winds (Fig. 6). In the most severe of these sites, limber pine occurs as the lone tree species, while on less severe sites it is mixed with Engelmann spruce and subalpine fir. In the southern half of the Front Range, bristlecone pine (*Pinus aristata*) also occurs on similar sites, either in pure stands or sharing dominance with limber pine (Peet 1978). Typically, limber pine forms open stands of well-branched and broad-crowned trees. The ground cover is usually sparse, with large patches of exposed, bare ground. The common understory species include many alpine species as well as shrubs such as common juniper, myrtle blueberry, and kinnikinnik.

Aspen stands occur over a wide variety of sites ranging from dry, high-elevation grasslands to poorly drained sites along the edges of wet meadows (Peet 1981). Like lodgepole pine, aspen forms extensive stands following burning of the previous forest. On xeric sites that could also be occupied by lodgepole pine, aspen tends to occur on sites with deeper and relatively fertile soils. Most aspen stands consist of clones of stems that have resulted from root suckering (i.e., vegetative reproduction). Although aspen stands are quite heterogeneous in their characteristics, common features are their open tree canopies and the relatively high light levels of the understory. In comparison with adjacent conifer stands, the understories of aspen stands are typically lush and rich in herbaceous species. Often seedlings and saplings of species such as Engelmann spruce, subalpine fir, and lodgepole pine occur in these understories.

Nonforest vegetation types of the subalpine zone include the willow and birch thickets of wet, marshy soils surrounding lakes. Several species of willow (*Salix* spp.) and the dwarf birch (*Betula glandulosa*) produce dense thickets 1 to 1.5 m tall. Flat sites too poorly drained for tree growth are characterized by wet meadows dominated by sedges. Restricted areas of dry subalpine meadow are also occasionally interspersed among the coniferous forests. These meadows are characterized by a great diversity of forbs.

Human Settlement

As will be quickly appreciated from a glance at some of the photographs of the vegetation of the Front Range, disturbance of the vegetation by humans has played a major role in shaping the present vegetation patterns. Thus, some consideration of both prehistoric and historic human occupation of the Front Range is essential to understanding present vegetation patterns.

Little is known about the details of the earliest human use of the Colorado Front Range. Thirty-five miles east of the mountains at the Dent archaeological site, remains have been found of early humans associated with mammoth remains dating from 9000 to 10,000 years B.C. (Buccholtz 1983). Archaeological remains of projectile points indicate the presence of campsites at tundra and timberline sites in the Front Range at least as early as 6000 years B.C. (Husted 1965; Benedict 1974). These campsites, such as those near Caribou Lake and on Trail Ridge, may reflect seasonal use of high altitude resources or travel between the plains and the mountain parks. Travel between the plains to the east and the high mountain parks (e.g., North Park and South Park) would have been easiest along several of the high, remnant erosion surfaces such as Trail Ridge (Fig.10). In summer and early fall, the tundra and subalpine habitats supplied forage for grazing animals such as elk, bison, antelope, and mountain sheep, which were hunted by the early Indians who lived the rest of the year in the foothills or on the plains. When the mammoths became extinct about 9000 B.C., due to climatic change or hunting pressure or a combination of both, bison became the source of subsistence for the nomadic hunters occupying the western Great Plains (Hughes 1977).

Abundant game drive fences in tundra habitats reflect the use of the high elevations of the Front Range for hunting over a period of at least 5800 years (Benedict 1975). More than 40 low-walled rock structures have been discovered near the crest of the Front Range. These are laid stone walls and rock cairns that produce a slight barrier on slopes that otherwise lack cover for game animals. Presumably, the rock walls aided in the driving of animals towards hidden hunters. Mule deer, mountain sheep, elk, and bison are believed to have been the primary animals hunted (Buccholtz 1983). The presence of a hunting economy in the Front Range implies that the native inhabitants were also setting fires for the purpose of driving game, as observed on the plains by early European visitors (Stewart 1956). However, the frequency of Indian-set fires relative to those ignited by lightning strikes is not known. Dated artifacts, ranging from fire pits to grinding tools, indicate increased visitation of the Front Range from 2500 B.C. onward.

By about 1000 years ago the Utes, who were hunters and gatherers, were present in the Front Range. They wandered over extensive areas from the plains to the mountains. The Utes and the Shoshoni were probably the most abundant groups using the higher elevations of the Front Range prior to the nineteenth century. In the 1790s the Arapahos arrived in the western plains, having migrated from the Red River country and Manitoba far to the northeast. Although the plains were their heartland, they also hunted extensively in the mountains (Buccholtz 1983). Other groups that roamed the adjacent plains and the foothills of the Front Range include the Comanche, Kiowa, and Cheyenne.

Although 1859 is the earliest date of confirmed European presence in the Front Range, there are possibilities of earlier undocumented brief visits by the Spaniards or French (Buccholtz 1983). When the Spanish began exploring what today is the southern part of Colorado in the eighteenth century, Indians told them of French traders coming from the East. While making an expedition to the Platte River in Nebraska in 1720, the Spanish are likely to have seen the Front Range, but they did not visit it. Early presence of the Spanish in the Front Range is suggested by an 1859 report by a prospector named Samuel Stone (Buccholtz 1983). Near the base of Longs Peak, Stone found the remains of an old mining camp including shafts, excavations, and cabins, leading to speculation about an early Spanish mining venture in the Front Range.

French and American fur traders in the eighteenth century probably visited the Front Range but have left no written reports. James Purcell was one of the first white American adventurers to wander into the Rockies. He traveled up the South Platte River and was captured temporarily by the Kiowas in South Park in 1803 (Smith 1981). In 1806 in Santa Fe (then under Spanish control) he told Zebulon Pike that he had discovered gold at the head of the Platte. Although Pike and his men spent several months traveling along the eastern edge of the Rockies, they did not visit the Front Range. Groups of Missouri beaver trappers worked the Front Range between 1811 and 1817, only to have their valuable collection of pelts confiscated by the Spaniards (Buccholtz 1983). In 1820 Major Stephen Harriman Long's expedition reached Colorado via the South Fork of the Platte. They were directed by three French guides to the base of the Rockies and traveled southward along the foothills of the Front Range. They visited Pike's Peak and then returned eastward.

For the three decades following the Long expedition, trappers continued to be the only white American visitors to the Front Range. Most of these mountain men lived in solitude or in some cases wintered with the Indians. They left no written accounts of their observations and are known primarily through their campfire conversations with more literate travelers of the West, such as George Ruxton (1951). The existence of a lucrative fur trade, undoubtedly including trapping in the Front Range, is indicated by the establishment of small trading posts along the South Platte River just east of the Front Range. Fort Vasquez was established in 1835 near the mouth of Cherry Creek and reestablished in 1838 near the present town of Platteville (Buccholtz 1983). About 1838, Ceran St. Vrain and his partners, William and Charles Bent, established Fort St. Vrain near the mouth of St. Vrain Creek. Also, by 1836 Lieutenant Lancaster P. Lupton had built a trading post originally called Fort Lancaster near the present town of Fort Lupton. In 1837, Henry Fraeb and Peter Sarpy built Fort Jackson about ten miles south of Fort St. Vrain (Buccholtz 1983). Initially all these trading posts competed for the trade in beaver pelts and buffalo hides. When men's hat fashions changed from beaver to silk, decline in demand for beaver pelt led to the abandonment of all four forts by the late 1840s.

One of the first white Americans to leave a written description of the Front Range was Rufus B. Sage, who traveled throughout the West between 1841 and 1844. From Fort Lancaster in 1843 he traveled "for ten or twelve miles, through a broad opening between two mountain ridges, bearing a northwesterly direction, to a large valley skirting a tributary of Thompson's creek, where finding an abundance of deer, I passed the interval till my return to the Fort." Some historians credit Sage as being the first white man to provide a written description of Estes Park on the eastern side of the present Rocky Mountain Park. Sage described the area as ". . . a concentration of beautiful lateral valleys, intersected by meandering watercourses, ridged by lofty ledges of precipitous rock, and hemmed in upon the west by vast piles of mountains climbing beyond the clouds, and upon the north, south and east, by sharp lines of hills that skirted the prairie . . ." (Hafen and Hafen 1956).

Although gold had been reported several

times during the first half of the nineteenth century, it was not until the 1858 gold find at Cherry Creek that a "gold rush" resulted. This was the beginning of permanent white settlement in the Front Range. William Green Russell led an expedition of Georgian miners to a site near the present location of Denver. They discovered gold in July 1858 along Cherry Creek at a site they named Auraria (Smith 1981). They sent samples of their find with a traveler to Kansas City and set off into the mountains to look for the source of the gold (Pettem 1980). News of the strike spread quickly from Kansas City and the gold rush to "Pike's Peak Country" was under way in 1858. By the time the Russell party returned from the mountains in the fall of 1858, cabins had been built on the adjacent plains, and in November Denver was founded (Pettem 1980).

Thomas A. Aikins, at the time a farmer in Missouri, heard of the easy riches to be found at Cherry Creek and set out with a small gold-seeking party via Nebraska City. Near Kearny, Nebraska, he merged his group with Alfred A. Brookfield's company and reached Fort St. Vrain in October of 1858. Rather than following the South Platte to Cherry Creek, they headed to the present location of Boulder (Pettem 1980). From the South Fork of the St. Vrain they headed up Boulder Creek to the mouth of Boulder Canyon (Fig. 7). After building about a dozen crude cabins, they quickly began prospecting and found their first gold in streams in the foothills in November 1858. In January 1859 they made a major find of a large placer deposit at Gold Run, 12 miles west of Red Rocks and 1000 m higher in elevation (Smith 1981). The Boulder City Town Company was officially formed in February 1859. About the same time important gold discoveries were made on the North Fork of Clear Creek between today's Central City and Black Hawk. Two weeks later an estimated 3000 miners were combing the area for gold (Buccholtz 1983). Another important find of gold in early 1859 was on Chicago Creek near Idaho Springs. By late winter of 1859 gold had been found in many of the streams of the Front Range, from the Arkansas River in the south to the Cache La Poudre in the north (Pettem 1980).

The Arapaho Indians who had been using Boulder Valley visited the Aikins camp occasionally. They quickly realized that the discovery of gold meant that they would be driven from their traditional hunting grounds. In 1860, the Arapaho held a final antelope hunt involving some 400 Indians near the site of Valmont (Smith 1981). The following year, a new treaty signed at Fort Lyon (near Bent's Fort on the Arkansas River) reduced the Arapaho hunting grounds by 90 percent. As the white settlers poured into the Front Range, the Indians visited less and less.

During 1859 and 1860 a dozen boom towns such as Denver, Boulder, and Golden mushroomed overnight. It is estimated that as many as 100,000 people came to Colorado as "Fifty-niners" (Buccholtz 1983). Many of the prospectors failed to make significant strikes, became frustrated, and returned to the East. Others, however, stayed and created permanent Front Range communities. In addition to the miners, there were the farmers, ranchers, merchants, saloon-keepers, and bankers.

One of the people attracted to Colorado by the gold was the Kentucky-born Joel Estes. He was in California when he heard of the Rocky Mountain gold and packed up his family to head for Colorado. After a brief stay in Golden in 1859, he settled 22 miles north of Denver near the South Platte River and developed a cattle ranch. In October, while on a hunting or prospecting trip to the mountains, he stumbled onto the valley of Estes Park. In 1860 he returned to Estes Park to claim the excellent cattle range and set up permanent residence. He hunted wild game and sold meat to prospectors too busy to take time to hunt for their supper.

Another early settler of the present Rocky Mountain Park area was Philip Crawshaw, who built a cabin about 1857 near Grand Lake (Buccholtz 1983). He trapped beaver along the North Fork of the Colorado and traded fur for gold dust in Denver. Later, hunting for elk was the dominant economic activity on the western side of the Front Range in the Grand Lake area. In 1875, following the discovery of silver at Wolverine, there was a short-lived rush to the Never Summer Range (Buccholtz 1983). An important silver discovery was made in 1879 at Shipler Mountain near the North Fork of the Colorado River. The boom town of Lulu City was quickly created but three years later both the mine and the town had declined.

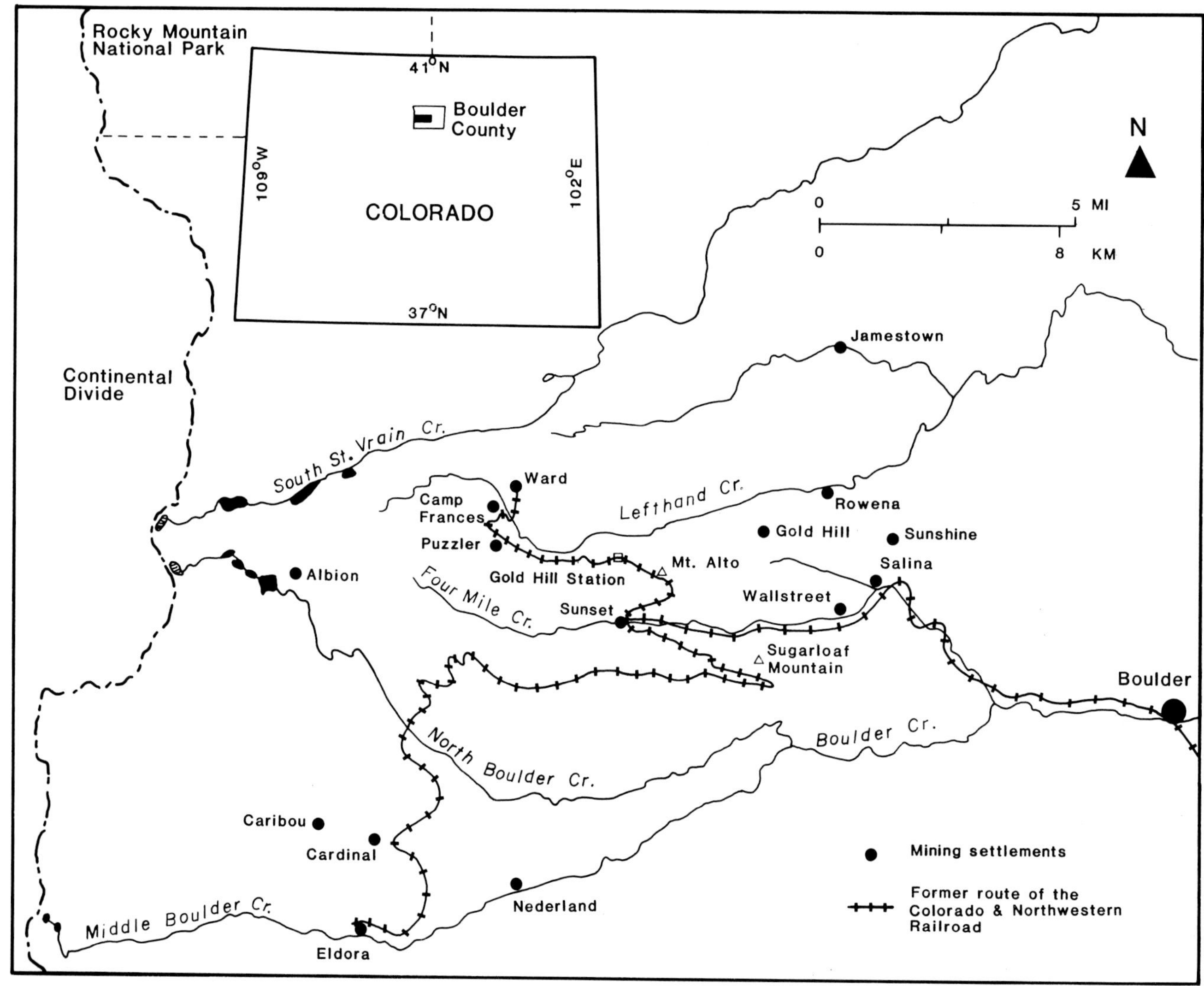

Figure 7. Historical map of the mining district of Boulder County.

In general, the northern part of the Front Range lies mostly outside the mineralized belt of the Rockies and consequently mining was of only minor significance in the area that was to become Rocky Mountain National Park. In contrast, in Boulder County and adjacent parts of the central and southern portion of the Front Range, mining flourished. By mid-1860, the mining town of Gold Hill supported 2000 residents (Smith 1981). Silver was discovered at Caribou in 1869 at 3200 m and within a year the town had nearly 500 residents (Fossett 1880). Tellurium, used primarily in the rubber industry as a vulcanizing agent, was discovered near Gold Hill in 1872 and initiated another rush (Wolle 1949). Other important boom-and-bust mining towns in the Front Range included Nederland, Ward, Cardinal, Eldora, Salina, Rowena, Rollinsville, Wallstreet, Sunset, and many others (Fig. 7).

With the intensive development of mining in Boulder County came the development of wagon roads and later of railroads. Early in the 1860s an important wagon road was constructed from the mouth of Left Hand Canyon to Ward, then south to Brown's Crossing (now Nederland), to Rollinsville, and on to Black Hawk (Kemp 1960). Roads for traversing the Front Range were constructed through Berthoud Pass and Rollins Pass. In 1865 a group of Mormons led by John Quincy Adams Rollins crossed the Continental Divide via Rollins Pass, or Boulder Pass as it was known then. They had to dismantle their 39 wagons and pack them piece by piece up the last part of the trail. Rollins subsequently constructed a wagon road, finished in 1873, for hauling freight over the pass that bears his name. Construction of the Moffat Railway began in 1903, and the railroad ran over Rollins Pass from 1904 to 1929. With the completion of the 10-km-long Moffat Tunnel through the Continental Divide in 1927, the "Hill Route" over Rollins Pass was no longer needed and the rails of the Moffat Railway were quickly dismantled.

In 1869, a wagon road up Boulder Canyon with 33 bridges was completed. The Union Pacific Railroad was constructed westward into the mountains from Boulder. By 1883 14 miles of narrow-gauge track connected Boulder to Salina, Wallstreet, and Sunset in Four Mile Canyon (Smith 1981). The devastating 1894 flood, which rushed through Four Mile Canyon and destroyed much of Boulder, ruined four miles of track, but within four years the line was rebuilt and extended to Ward (Crossen 1978). By 1904, a branch from Sunset to Eldora was built by the Colorado and Northwestern Railroad. Although built primarily to transport ore, the narrow gauge known as the Switzerland Trail was an important tourist attraction as well (Fig. 7). Due to the decline in mining activity and competition from trucking, the railroad ceased operations in 1919 and the Switzerland Trail was dismantled (Crossen 1978). In contrast to the abundant road and railroad construction in Boulder County, in the mineral-poor area of present Rocky Mountain National Park there was no railroad construction and only a few wagon roads were built during the latter part of the nineteenth century.

The mining booms of the central and southern part of the Front Range resulted in heavy demands on the timber resources for fuel, mine props, and town construction (Fritz 1933; Kemp 1960). The forests of Boulder County, particularly those of the montane zone, were heavily logged during the nineteenth-century mining booms. In many areas, such as Boulder and Four Mile Canyons, nearly all the timber was cut. Logging during the late nineteenth century in the northern part of the Front Range was much less extensive. Nevertheless, around the turn of the century a surprisingly large percentage of the forests in present Rocky Mountain National Park were logged for fuel and construction by ranchers and resort operators. Permits for logging were issued in the early years of the Park as if it were still a national forest (Buccholtz 1983).

During the last half of the nineteenth century, catastrophic fires had an equal, if not greater, impact than logging on the montane forests of the Front Range (Tice 1872; Fossett 1880; Fritz 1933; Wolle 1949; Kemp 1960). Many of these fires were set by humans, either accidentally or intentionally. Some fires were set "merely to clear away the fallen leaves so as to expose the naked rocks to the observation of the prospector" (Tice 1872:123). Although the mining districts outlawed forest burning and imposed severe fines, the burning continued. In 1871 in Boulder County, there were 51 indictments for illegal fires (Tice 1872). In 1860, a forest fire raged through Gold Hill, destroying the recently founded town (Wolle 1949). The town was nearly destroyed again in 1894 by a large forest fire that started near Ward (Smith 1981). In 1873 and 1879, Caribou was razed by fires (Kemp 1960; Smith 1981). About 1900, a single fire burned an estimated 29,000 ha of forest in the area of Eldora in southwestern Boulder County (Kemp 1960). Since the 1920s, fire suppression has greatly decreased the frequency and extent of forest fires.

Much of the Front Range was included in the Medicine Bow Forest Reserve in 1905. The Colorado section of this reserve became the Colorado Forest Reserve in 1910 and in 1932 was renamed the Roosevelt National Forest, which today includes much of the mountainous portion of Boulder County. Logging, particularly of the subalpine forests, was important in Roosevelt National Forest until the 1960s. Since then, recreational use has predominated and in recent years summer visitors have numbered in the hundreds of thousands. In 1915, Rocky Mountain National Park, encompassing most of the Larimer County portion of the Front Range, was created, and in recent years it has attracted many visitors from around the nation and the world.

Forest Ecology of the Front Range

The Autecology of Front Range Trees

To understand the patterns and processes of vegetation change in the Front Range, one needs information about both the ecology of the major tree species and the history and characteristics of the disturbances that have affected the area. Autecology is the study of the responses of an individual organism, in this case a tree, to its environment. An understanding of an organism's autecology is a fundamental first step in studying the ecology of communities in which that organism occurs. The ability of the major tree species of the Front Range to tolerate varying intensities of environmental factors such as moisture, temperature, light levels, and nutrient availability are largely inferred from the environmental conditions of the sites where they occur (e.g., Marr 1961; Peet 1981) and to a lesser extent from experimental evidence (e.g., Knapp and Smith 1981, 1982).

Ponderosa pine is a drought-resistant species but it is not tolerant of shade. It is substantially less shade tolerant than Douglas fir, its principal associate in the Front Range. It has a long taproot, which is advantageous both in obtaining adequate moisture and in decreasing the likelihood of uprooting by the strong chinook winds that frequently buffet the forests of the foothills. Ponderosa pine is a long-lived species capable of attaining ages well over 500 years (Keen 1940). In competition with grasses, the establishment of ponderosa pine seedlings is often limited to sites of stony, coarse-textured soils. On more fine-textured soils, moisture infiltration is slower and gives an advantage to the grasses because of their shallow but dense networks of roots. Grass roots generally grow faster and deplete soil moisture to lower levels than ponderosa pine (Schubert 1974). So, ponderosa pine seedlings grow and survive better in the absence of grasses. Due to its thick bark, ponderosa pine is the most fire-resistant species of the Front Range. Trees greater than 15 cm in diameter often survive the light surface fires characteristic of open ponderosa pine woodlands. Thus, low intensity surface fires kill seedlings and saplings but mature trees survive fire.

Dwarf mistletoe (*Arceuthobium vaginatum*) is a serious parasite affecting ponderosa pine in the Front Range. It is present on about 20 percent of the ponderosa pine forests of the Front Range (Alexander 1986). Dwarf mistletoe reduces seed production and growth rates and contributes to tree death. Young trees are easily killed by the parasite whereas large trees with well-developed crowns may sustain an attack for many years without showing signs of ill health. Severely damaged stands have many trees with deformed branch growth, known as "witches' brooms," and dead tops or "spiketops."

Many species of insects infest ponderosa pine (Stevens et al. 1980) but the mountain pine beetle (*Dendroctonus ponderosae*) is by far the most serious insect pest. The beetles bore through the bark, create egg galleries, mate, and deposit eggs in the phloem layer. They carry with them blue-stain fungi, which in conjunction with the beetle block water and nutrient-conducting tissues and girdle the tree, causing its death. Epidemic outbreaks are most likely in stands in which most of the trees are greater than 15 cm in diameter and which are overcrowded and under stress (Alexander 1986). Natural factors that cause the decline of mountain pine beetle outbreaks include unusually cold temperatures in fall or winter, parasite attacks on the beetles, predation by woodpeckers, and simply exhaustion of the food resources as the trees die (McCambridge and Trostle 1972). Other insects that attack ponderosa pine in the Front Range include the red turpentine beetle (*Dendroctonus valens*) and pine engraver beetles (*Ips*), but both are relatively minor in comparison with the mountain pine beetle (Alexander 1986).

Compared to ponderosa pine, Douglas fir is substantially less tolerant of xeric conditions but more tolerant of shade. Thus, the establishment of Douglas fir seedlings is common in the understories of stands dominated by ponderosa pine and/or Douglas fir. Douglas fir seedlings will also establish under open conditions, but on dry sites they are more susceptible to desiccation than are the seedlings of ponderosa pine. The thick bark of mature trees allows them to survive light surface fires that kill seedlings and saplings.

The two major insect pests of Douglas fir in the Front Range are the western spruce budworm (*Choristoneura occidentalis*) and Douglas fir bark beetle (*Dendroctonus pseudotsugae*). Epidemics of spruce budworm appear to be triggered by warm, dry weather and may last for 15 years or more (Sheppard 1959; Campbell 1986). Spruce budworm is a defoliator which in the Front Range feeds mainly on Douglas fir and secondarily on Engelmann spruce and blue spruce, where the latter two species descend to lower elevations along streams. Larvae feed as typical defoliators, preferring expanding buds and succulent new foliage, and as the new foliage is eliminated, they feed also on the older foliage. They feed on Douglas fir cones and seeds as well (Carlson et al. 1982). Although insect parasites and both vertebrate and invertebrate predators may influence budworm populations, adverse weather appears to be the most important natural factor controlling epidemics. Cool summers, late spring frosts, and wet summers tend to retard larval development and feeding. Starvation is also an important cause of mortality when stands are heavily defoliated (Carlson et al. 1982). The amount of tree mortality caused by budworm is highly variable, depending on the intensity and duration of the attack and the size and vigor of host trees. Smaller-diameter understory individuals suffer the greatest immediate mortality from defoliation. Large-diameter canopy trees are more likely to recover from defoliation but in their weakened state often become more susceptible to bark beetle attacks (Cates and Alexander 1982).

In addition to predisposition of trees to Douglas fir bark-beetle attack by prior budworm defoliation, damage by fire, drought, air pollution, or logging may also increase susceptibility to beetle attack (Cates and Alexander 1982). Bark beetles prefer large-diameter host trees with thick phloem, which provide a reliable food supply. Dwarf mistletoes are also parasitic on Douglas fir and often weaken trees, making them more susceptible to other damaging agents.

Lodgepole pine is highly shade intolerant but capable of surviving on xeric sites (Alexander 1974; Knapp and Smith 1981). It is shallow rooted and highly susceptible to windthrow. Lodgepole pine is often considered a classic example of a species adapted to wildfire. Although it is thin barked, so that even mature trees are not fire resistant, it has long been recognized as a species capable of vigorous new establishment following fire (Clements 1910). Lodgepole pine possesses cones that remain closed and attached to the tree for many years. Cone production begins at a very early age, as young as 10 years, and large numbers of mostly closed cones accumulate on standing trees. Following a stand-devastating fire, these cones, which are opened by the heat of the fire, release a large number of seeds, which in theory will germinate and establish in the seedbed of ash left by the fire. At some sites, however, herbs may form an extensive cover after fire, and lodgepole pine establishment may be exceedingly slow. On most open sites, lodgepole pine is capable of substantially more rapid growth than its more shade-tolerant associates such as Douglas fir at low elevations and Engelmann spruce and subalpine fir at higher elevations. Thus, intense fires generally favor the establishment of a new stand dominated by lodgepole pine. However, there is a great deal of variability in the percentage of trees with nonserotinous cones according to site conditions and stand history (Lotan 1975; Muir and Lotan 1985). For example, in stands that originated following large-scale blowdown or lethal insect attack, a small percentage of the cones may be serotinous. This may result in an inadequate supply of seed for regeneration following a fire (Muir and Lotan 1985; Alexander 1974).

Dwarf mistletoe, lodgepole pine beetle (*Dendroctonus murrayanae*), and mountain pine beetle are the most important natural biotic agents that damage lodgepole pine stands. Dwarf mistletoe reduces growth rates and makes trees more susceptible to other agents of mortality. Lodgepole and mountain pine beetle outbreaks usually reach epidemic proportions only in stands that contain at least some trees greater than 30 cm in diameter (Alexander 1974). Trees less than 15 cm in diameter are rarely killed (Cole and Amman 1969). Although mountain pine beetle has caused devastating epidemics in lodgepole forests elsewhere in Colorado, no major epidemics of this insect or of lodgepole pine beetle have affected the lodgepole pine forests of the Front Range (Barrows 1936; Roe and Amman 1970). Epidemics of this insect in the Front Range have, however, greatly affected ponderosa pine.

Engelmann spruce trees in the Front Range often attain ages of over 300 years, and trees 500 to 600 years old are not uncommon (Peet 1981; Veblen 1986a). It is substantially more shade tolerant than lodgepole pine and limber pine, so that seedling establishment is common beneath a forest cover. Under shaded conditions its seedlings are typically much less abundant than those of subalpine fir, but both species occur under similarly low levels of light (Knapp and Smith 1982). Seedling establishment is generally better on bare mineral soil than on undisturbed forest litter (Alexander and Shepperd 1984). In old-growth stands in relatively mesic habitats a large percentage of the seedlings occur on moss-covered decaying logs. Where the water table is near the surface or on soils underlain by impervious rock or hardpans, the root system is shallow and trees are highly susceptible to windthrow. The thin bark of Engelmann spruce makes this a highly fire-sensitive species. Persistence of dead lower limbs increases the likelihood of intense crown fires and of tree death.

The most common diseases affecting Engelmann spruce are woodrotting fungi, which predispose trees to windthrow and windsnap (Hinds and Hawksworth 1966). Spruce broom rust (*Chrysomyxa arctostaphyli*) is common and causes bole deformation, reduced growth, and spiketops. The spruce beetle (*Dendroctonus rufipennis*), which kills spruce in much the same way that mountain pine beetles kill ponderosa pine, is its most serious insect pest. Endemic (i.e., always present but at low densities) beetle populations contribute some mortality in most old-growth stands, and where blowdown or logging leave ample food supplies for population buildups, extensive epidemics may occur. Outbreaks have caused severe damage, including mortality of well over 50 percent of the basal area of spruce in many parts of the subalpine zone of Colorado, but no major outbreaks have been recorded in this century in the Front Range (Schmid and Frye 1977). Large-diameter trees are preferentially attacked, but if an infestation persists, smaller trees will also be attacked and killed. Epidemics are terminated either by extremely low temperatures or by elimination of the insects' food supply (Schmid and Beckwith 1972).

Subalpine fir characteristically grows in association with Engelmann spruce in the subalpine zone. Although trees older than 250 years are not uncommon, this species rarely reaches ages greater than 350 years. Subalpine fir is a relatively shade-tolerant species. Ecophysiological studies have documented its greater photosynthetic capacity at low light levels compared to Engelmann spruce (Knapp and Smith 1982). It also has a more rapidly growing and deeper taproot, which enables it to penetrate the rapidly drying litter layer of mature spruce-fir forests. Thus, seedlings of subalpine fir are consistently much more abundant in forest understories than those of Engelmann spruce (Whipple and Dix 1979; Peet 1981; Veblen 1986b). In the krummholz zone, both fir and spruce commonly reproduce vegetatively by layering (i.e., the rooting of branch tips). Even in tall forests, both species sometimes reproduce by layering of low branches or basal branching (Shea and Grant 1986).

Like most of the other conifers of the subalpine zone, subalpine fir is shallow rooted and easily subject to windfall (Alexander et al. 1984). Its thin bark and the persistence of dead lower limbs make it particularly susceptible to being killed by fire. A variety of wood-rotting fungi infect subalpine fir even at relatively early tree ages. These fungi make trees susceptible to windthrow and windsnap and account for the relative lack of old individuals of this species. The most destructive insects attacking subalpine fir are the western balsam bark beetle (*Dryocoetes confusus*) and the fir engraver (*Scolytus ventralis*). Both can cause substantial local mortality, but there are no reports of massive epidemics of these insects in the Front Range (Furniss and Carolin 1977). Western spruce budworm also attacks subalpine fir, but primarily only in the lower part of its elevational range.

Limber pine is the least shade tolerant of the conifers occurring in the Front Range. It has a deep taproot and therefore is the most windfirm tree species of the zone. It is thin barked and consequently easily killed by fire but may be able to tolerate fire slightly better than Engelmann spruce and subalpine fir. It is capable of attaining great ages, and trees over 700 years old are common in the Front Range. In the lower part of its elevational range, it may be attacked by mountain pine beetles, but in general it suffers from fewer potentially lethal inspect pests than the other conifers.

Aspen typically is a short-lived tree, rarely attaining ages greater than 120 years. It is highly susceptible to a canker disease that is easily spread where the thin bark of aspen is damaged by animals or humans (Hinds 1976). Due to its relatively weak stems, aspen is often severely damaged by heavy snow accumulations and avalanches. It is easily killed by fire but vigorously root-suckers following fire (Marr 1961).

Disturbance History of the Front Range

Although detailed information on the history of disturbances affecting the vegetation of the Front Range is not available, it is possible to roughly outline some of the changes in the relative importance of disturbances over the past century or two. Partial information is available only for the major disturbances of fire, logging, and insect epidemics. Smaller-scale disturbances, such as windthrow, cannot be easily documented despite their obvious importance in some forest types.

The pervasive influence of fire on the forests of the Front Range is reflected by the ubiquitous presence of charcoal beneath forests over the entire elevational gradient (Peet 1981). There have been only a few attempts to quantify the changes in fire frequencies over the past few centuries in the Colorado Front Range. Such studies are based primarily on aging fire scars on trees that have survived past fires and to a lesser extent on the ages of cohorts (groups of similarly aged trees) that established following a stand-devastating fire. The results of fire-history studies must be interpreted cautiously because of the likelihood of significant variations in fire frequency over short distances due to sudden changes in site and forest conditions in the mountainous terrain. Although the available studies are not adequate for fully quantifying fire history in the Front Range, they do identify some consistent trends.

A useful way of describing fire frequency is on the basis of "mean fire-return interval," which is the average time between successive fires in a designated area. Prior to Euro-American settlement in the mid-nineteenth century, fires were ignited by either lightning or Indians. Both these sources of ignition are arbitrarily regarded as "natural" fires by fire ecologists. During the pre–Euro-American settlement era, as is also true today, there was a dramatic decrease in fire frequency as one goes from low to high elevations. Highest fire frequencies were in the open ponderosa pine woodlands at the boundary of the foothills with the plains. Although quantitative studies of fire frequencies in this vegetation type are lacking for the Front Range, studies elsewhere in the Rockies suggest that mean fire-return intervals were on the order of 5 to 12 years (Peet 1988). In the foothills where ponderosa pine forms denser stands mixed with Douglas fir on the more mesic sites, pre–Euro-American settlement mean fire-return intervals have been estimated at 39 to 66 years (Rowdabaugh 1978; Laven et al. 1980). Although there have not been any fire-history studies conducted in subalpine forests specifically in the Front Range, Romme and Knight (1981) have estimated mean fire-return intervals in Engelmann spruce–subalpine fir forests in the adjacent Medicine Bow mountains of southeastern Wyoming at 300 to 400 years. A similarly long fire interval has been documented for lodgepole pine–dominated forests in Yellowstone (Romme 1982). The low fire frequency of the subalpine zone mainly reflects the long time required for fuels to accumulate.

In addition to having very different frequencies, fires in the montane and subalpine zone differ significantly in average intensities. Most fires occurring in the open ponderosa pine woodlands are surface fires carried mainly by grass fuels and are unable to develop into crown fires due to the low density of trees and the lack of lower limbs on ponderosa pine. At higher elevations, in denser stands of ponderosa pine and particularly in mixed stands with Douglas fir, fires are more likely to become crown fires and are very patchy in spatial pattern as a result of the heterogeneous site and fuel conditions. Throughout the montane zone, most fires are light surface fires, but the infrequent crown fires have had a major impact on the landscape (Rowdabaugh 1978). In the subalpine zone, fires occur infrequently but are primarily crown fires that devastate extensive areas and initiate new stand development. Towards the alpine tundra in the timberline zone, fires again become patchy in their spatial pattern (Crane 1982). Although the pattern of frequent, low-intensity fires at low elevations and infrequent, high-intensity fires at high elevations is clear, there is substantial local variation. For example, where lodgepole pine or Engelmann spruce–subalpine fir forests are adjacent to grasslands or ponderosa pine woodlands, these subalpine forest types may be affected by low-intensity surface fires (Skinner and Laven 1983; Peet 1988).

Variation in fire frequencies related to white settlement follows a consistent pattern in the Front Range (Rowdabaugh 1978; Laven et al. 1980; Skinner and Laven 1983). During the settlement period from the mid-nineteenth century to about 1915 there was a dramatic increase in fire frequency. As described earlier, much of this increase resulted from intentional burning by prospectors. For the montane zone, fire frequency increased severalfold so that mean fire-return intervals decreased to less than 20 years. With effective fire suppression beginning in the 1920s, fire frequency declined dramatically. In the subalpine zone, white settlement also resulted in an increase in burning. This is particularly evident in the case of the lower part of the subalpine zone, where lodgepole stands approximately 100 years old are very extensive. Increased burning in the higher parts of the subalpine zone, including stands near timberline, may also have occurred but was substantially less extensive and resulted in a patchy pattern of burns. Due to the long natural fire interval in the subalpine zone, modern fire suppression has had a much less dramatic impact than at lower elevations.

Lethal insect outbreaks have been documented by Forest Service and National Park records and by tree-ring analyses. During the present century in the montane zone, there have been several outbreaks of insects such as western spruce budworm, Douglas fir bark beetle, and mountain pine beetle. Major epidemics of western spruce budworm affecting principally Douglas fir have occurred in the early 1940s, early 1960s, and early 1980s (Fig. 8). Although the timing of these outbreaks has varied from site to site, in most localities they lasted about six to nine years. Each stand may not not have been affected by all three outbreaks but most were affected by more than one outbreak. Tree-ring evidence indicates that budworm epidemics also occurred in the eighteenth and nineteenth centuries (Swetnam 1987). The frequency and intensity of budworm outbreaks appear to have increased during the past 40 years. Too few stands, however, have been examined to be certain that this is a real trend.

Douglas fir has also suffered from recent massive outbreaks of bark beetles. Outbreaks of this lethal insect occurred in the Front Range in the early 1950s and the mid-1980s. Typically, they follow budworm outbreaks. Prior to the 1950s, Douglas fir bark beetle was not regarded as a major problem in the Front Range (Barrows 1936).

Mountain pine beetle epidemics devastated ponderosa pine populations in the Front Range in the mid-1920s, mid-1930s, late 1950s, and mid-1970s. Although this beetle also attacks lodgepole pine, there have not been any recorded epidemic attacks on this species in the Front Range (Barrows 1936). This may reflect the fact that most of the lodgepole pine stands are relatively young stands that initiated during the settlement era and conse-

Figure 8. A mixed stand of Douglas fir, ponderosa pine, and aspen in which most of the Douglas fir have been defoliated by spruce budworm.

quently have too few trees large enough to support a major outbreak (Veblen and Lorenz 1986). Engraver beetles (*Ips* spp) have caused more damage to lodgepole pines than have mountain pine beetles, but apparently there have not been any region-wide outbreaks in the Front Range. Similarly, although spruce bark beetles are endemic in the spruce-fir forests of the Front Range, there are no recorded major outbreaks.

Extensive areas of timber were cut during the mining era beginning in the 1860s in the Front Range. The demand for fuel, mine timbers, and lumber resulted in clearcutting of the forests in much of the montane zone of Boulder County. Even in the area that today is Rocky Mountain National Park there was extensive cutting in the montane zone by pioneer ranchers and later by resort operators. Selective logging of spruce-fir forests affected a large part of the more accessible subalpine forests of Rocky Mountain National Park during the late nineteenth and early twentieth centuries. Since the early 1960s there has been relatively little cutting in Front Range forests. Much of the recent cutting has been for firewood or for control of insect outbreaks and has been supervised by the Forest Service.

Stand Development Patterns

Due to the widespread influence of fire in the Rockies, most studies of stand development have examined the structural and compositional changes that occur following fire (e.g., Whipple and Dix 1979; Peet 1981; Veblen and Lorenz 1986). Much less is known about stand responses to disturbance by blowdown, insect disturbance, or logging. An important conclusion to emerge from the many studies of stand development in the Colorado Rockies is the high degree of variation in patterns of stand development due to the existence of steep environmental gradients. Thus, over distances of just a few tens of meters, changes in aspect or topographic position may substantially alter patterns of succession and stand development (Peet 1981).

Below about 2000 m on gently rolling topography, steady-state stands of ponderosa pine are common (Peet 1981). In such stands the abundance of young trees is sufficient to assure replacement of the old individuals as they die. On more mesic sites, where Douglas fir occurs in association with ponderosa pine, the former species tends to replace successionally the latter. Following crown fires, both species may establish even-aged (i.e., trees all of a similar age) populations, but as the stand continues to age, Douglas fir regenerates more successfully than does ponderosa pine (Peet 1981; Veblen and Lorenz 1986). However, due to the extent of wildfire in the settlement era, ponderosa pine–Douglas fir stands in the Front Range are not old enough to have attained steady-state (Peet 1981). Where light surface fires frequently burn through the understory of a ponderosa pine and Douglas fir stand, they are likely to prevent this successional replacement by eliminating the small fire-sensitive Douglas fir. In the montane zone, limber pine also plays the role of a seral species being replaced by Douglas fir in the montane zone.

Insect epidemics in the montane zone clearly influence the patterns of stand development that have been initiated by stand-devastating fire. By killing the large ponderosa pine, mountain pine beetle outbreaks enhance the growth opportunities for Douglas fir, thereby accelerating the postfire successional trend. By causing more mortality among the small, suppressed Douglas fir, spruce budworm defoliation is likely to retard the successional replacement of pine by Douglas fir. Where Douglas fir beetle attack causes severe mortality, a shift in the relative dominance towards ponderosa pine will occur.

At higher elevations, in the upper montane zone and the lower part of the subalpine zone, stand-devastating fires are most likely to result in new stands dominated by lodgepole pine. Where trees with serotinous cones provide adequate seed sources, dense stands of lodgepole pine establish (e.g., Plates 1, 26, and 54). On favorable sites, such as mesic slopes in the middle of the species' elevational range, stands undergo initially rapid growth, followed by slower growth and rapid thinning until the next fire starts the cycle over. On less favorable sites or where seed is not initially available in sufficient quantities, seedling establishment and growth may be much slower, so that tree recruitment occurs over 30 to 50 years (Veblen and Lorenz 1986). On extremely poor sites, ''dog-hair'' stands develop in which all the trees grow very slowly, with little variation in tree size, and with extremely slow natural thinning. Towards its lower elevational range, lodgepole pine is successionally replaced by Douglas fir and at higher elevations by Engelmann spruce and subalpine fir (Marr 1961; Peet 1981). Although lodgepole pine is clearly a seral species on most sites in the Front Range, over a restricted elevational and moisture range where the shade-tolerant conifers are absent, lodgepole pine can form self-perpetuating stands (Peet 1981).

At higher elevations in the subalpine zone on moderately xeric sites, postfire colonization is initially by lodgepole pine, either alone or with varying numbers of spruce and subalpine fir (e.g., Plate 58). Establishment of lodgepole pine at these higher elevation sites is slow, often lasting well over 100 years (Whipple and Dix 1979; Veblen 1986a). Spruce may also establish in abundance soon after the fire and in some cases establishes earlier than lodgepole pine (Veblen 1986a). Where seed sources are available and where the site is less xeric, spruce is more likely to play a significant role in the early phases of stand development. In approximately 200-year-old stands, as the canopy cover closes, seedling establishment of lodgepole pine typically ceases and that of spruce is drastically reduced. Fir, in contrast, typically does not begin to establish abundantly for at least 50 years following stand initiation but continues to establish abundantly throughout the remainder of stand development (Whipple and Dix 1979; Veblen 1986a). As self-thinning occurs in the postfire cohort of lodgepole pine, dominance steadily shifts towards spruce and fir until the pines are entirely eliminated. Where blowdown occurs in subalpine lodgepole pine–dominated stands, it tends to accelerate the replacement of pine by the more shade-tolerant Engelmann spruce and subalpine fir. Due to their abundance as small understory individuals, the shade-tolerant species survive the blowdown better and increase their growth rates in response to the reduced competition from the canopy trees (Veblen et al. 1989). On more mesic sites, spruce tends to dominate, while on drier sites subalpine fir may be the principal species replacing the pines (Peet 1981).

At higher elevations and on more mesic sites, lodgepole pine is typically absent and Engelmann spruce is the principal pioneer species after fire (e.g., Plate 62). As these spruce-dominated stands develop, the abundance of subalpine fir steadily increases and the establishment of spruce becomes scarce. In old-growth stands with trees greater than 300 years old, subalpine fir seedlings and small trees (i.e., less than 8 cm diameter) are generally several times more abundant than those of spruce. In contrast, spruce is generally more abundant among the canopy trees (i.e., trees greater than 20 cm diameter) and is clearly dominant in basal area. The greater success of fir at establishing under low-light levels and on forest litter is apparently balanced by the much greater longevity of spruce. Thus, despite the greater abundance of the younger fir, the two species continue to coexist in the same stands and form the steady-state forests of the subalpine zone (Veblen 1986b).

At the most extreme sites of thin, rocky soils exposed to strong winds, limber pine typically dominates stand development following the burning of a subalpine forest (Peet 1981; Veblen 1986a). Limber pine may be the sole dominant for several decades or it may co-colonize the site with spruce, which along with fir eventually replaces the pine (e.g., Plate 34). On exceptionally harsh sites too unfavorable for spruce or fir, limber pine may form open stands of self-perpetuating populations (Marr 1961).

Aspen is similar to lodgepole pine in its dominance of postfire stands. Following fire, aspen produces abundant, rapidly growing root suckers so that burned stands that initially contained only rare aspen are likely to be replaced by extensive clones of this species (e.g., Plate 27). Aspen stands typically have open canopies beneath which light levels are relatively high. Thus, spruce, subalpine fir, and even lodgepole pine are able to establish and grow up through the aspen. The aspen shoots are short-lived, rarely living much longer than 100 years, so that the conifers quickly replace them (e.g., Plate 1). Where fires are intense, aspen roots may be killed, which will favor the initial development of a conifer-dominated stand (Parker and Parker 1983). However, aspen stands often have a well-developed understory of moist forbs and grasses, which results in relatively cool fires in comparison with the hot fires of the adjacent conifer stands with their dry understories and substantial accumulations of dry fuels. Thus, in aspen stands on mesic sites, relatively cool but stand-devastating fires favor the regeneration of aspen over that of conifers.

The Method of Repeat Photography

Repeat photography is the practice of finding the location of an earlier photograph, reoccupying the original camera position, and taking a new photograph of the same scene (Rogers et al. 1984). Repeat photography is used in a wide range of disciplines including geology, physical geography, botany, archaeology, and the social sciences. Comparing photographs of the same landscape taken many years apart provides an effective means of investigating changes in vegetation and has been widely used by ecologists and geographers (Hastings and Turner 1965; Gruell 1980; Rogers 1982; Vale 1987; Veblen and Lorenz 1988).

The photographs shown here are from a variety of sources. A major and certainly the most convenient source of historical photographs is the Photographic Library of the U.S. Geological Survey in Denver. This library contains a collection of approximately 250,000 photographs of subjects taken during geologic explorations of the United States from 1867 to the present. The photographs are indexed by subject and by geographic location, making the collection convenient to use. Most of the photographs were taken to illustrate geologic and physiographic features, but many were also taken specifically to document vegetation conditions. Some photographs are succinctly annotated as to location, subject, and date, but many lack precise locations and dates. For photographs lacking dates, approximate dates may be determined by knowing the date of a particular expedition or the time of professional activity of the photographer. Another photographic collection that proved useful for this study was in the Western History Department of the Denver Public Library. This is the repository of the photographic collection of the famous Coloradan photographer Louis Charles McClure. The third principal source of photographs was the Boulder Historical Society, which houses the M. R. Parsons collection of Joseph B. Sturtevant photographs. Additional minor sources of photographs include the Rocky Mountain National Park Historical Collection and the Western History Collection of Norlin Libary at the University of Colorado in Boulder.

The several hundred photographs considered in this study were taken by numerous photographers among whom three individuals stand out due to the early dates and high quality of their landscape photographs. The most famous pioneer photographer of the Rocky Mountain region is William Henry Jackson (1843–1942). He was the official photographer of the 1870s Hayden Survey of the Teton and Wind River Ranges of Wyoming and Colorado. He later settled in Denver, where he opened a photographic studio. Louis Charles McClure (1867–1957) was trained by William Henry Jackson in the 1880s. When Jackson sold his studio and moved to Detroit in 1897, McClure remained in Denver (Jones and Jones 1983). The objects of McClure's photography were highly varied, including architecture, railroads, and cityscapes as well as mountainous landscapes. He produced spectacular landscape photographs over an extensive area of the Colorado Rockies

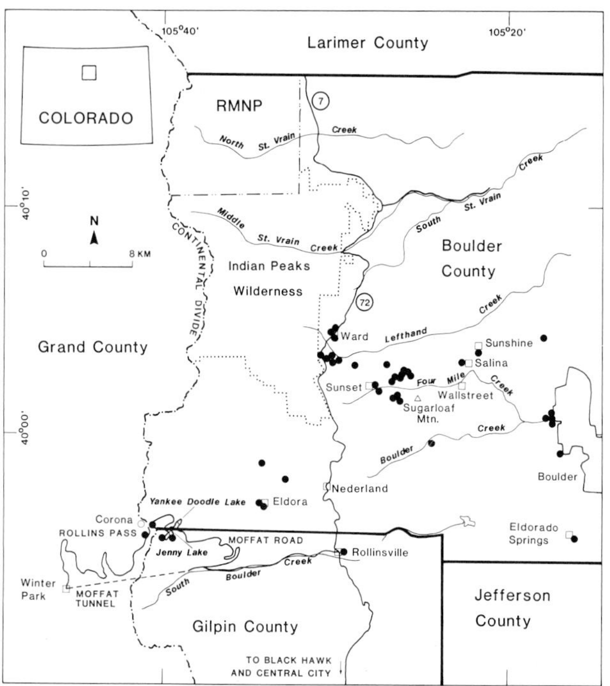

Figure 9. Map of Boulder County and adjacent areas. Solid circles indicate camera points of matched photographs. Empty circles indicate former settlements. Empty squares indicate present settlements.

Figure 10. Map of Larimer County and adjacent areas. Solid circles indicate camera points of matched photographs. Empty circles indicate former settlements. Empty squares indicate present settlements.

primarily during the period from 1900 to 1920. Joseph B. Sturtevant (1851–1910) was a pioneer Boulder artist and photographer, popularly known as "Rocky Mountain Joe" (Schoolland 1982). His photographic objects were also varied but he had a particular interest in transportation and mining. His photographs of the Front Range date from 1870 to 1909.

In our study of vegetation change in the Front Range, we relocated and rephotographed approximately 120 historical photographs, but only 69 matches are reproduced here. The locations of the photographs are spread over broad elevational and north-to-south ranges (Figs. 9 and 10). The historical photographs were taken mostly between 1880 and 1915 and the contemporary photographs between 1984 and 1986. Methods for matching old photographs are described by Harrison (1974), Malde (1973), and Rogers et al. (1984). Even when the location of a photograph has been noted by the original photographer, it still requires considerable time (often several hours) to relocate exactly the camera point. In some cases, due to a lack of easily identifiable landscape features, relocation may be nearly impossible.

Where detailed measurements are to be made on the photographs an exact match is essential. This requires not only finding the exact camera location for the original photograph but also reproducing the height of the camera point, camera format, and lens characteristics (Malde 1973). In many cases, however, it may be impossible or impractical to produce an exact match. For example, many original camera points were along early wagon roads that have since been drastically modified by highway construction. Occasionally, an exact relocation would have required standing on a three-meter ladder in the center of a busy highway. Excavation for building or reservoir construction has destroyed some former camera points. In other cases, the view from the original camera point may now be entirely obscured by trees. Since our objective was to qualitatively compare landscape scenes at two points in time (instead of quantitative measurements), we accepted approximate matches. We did not attempt to reproduce the format of the box cameras used by the early photographers and instead used a 35-mm Nikon camera. We did, however, attempt to approximate the original focal length by adjusting a zoom Nikon 28-to-90 mm lens. Further matching was accomplished in the printing process by cropping the old and new photographs to the same size. Although the comparison of photographed scenes in our study is largely qualitative, considerable supplemental quantitative data were collected on the forest stands themselves (Veblen and Lorenz 1986). At each rephotographed scene, precise locations were noted and detailed observations on forest composition and condition, including evidence of disturbance, were made. Within the scenes of some historical photographs quantitative samples of the vegetation were taken. In 24 forest stands in the historical photographs rectangular plots (from 250 to 1000 m in size) were located, and all trees were measured and aged. Plot sizes were selected according to the density of the stand so that approximately 50 trees per stand were sampled. Aging of trees was accomplished by taking increment core samples near the base of the tree, on which annual rings were counted after carefully preparing the surfaces of the samples by sanding or cutting with a sharp blade. For seedlings and saplings too small to core, counts of their total populations were made. The diameters of dead standing trees and cut stumps were measured. Where feasible, bark and wood characteristics were used to identify the species of these dead individuals. A detailed interpretation of these data is available in Veblen and Lorenz (1986) and will not be repeated here. However, many of the interpretations of the photographic comparisons presented here are derived with the aid of the quantitative data in this study.

The Plates

For photographs not dated by the photographer, estimated dates (e.g., c. 1910) are derived from consideration of the time of the photographer's professional activity in Colorado and dated changes in the photograph (e.g., road or building construction). The following abbreviations are used for the sources of the photographs: DPL, Denver Public Library; BHS, Boulder Historical Society; USGS, U.S. Geological Survey Photographic Library (Denver); CHS, Colorado Historical Society; CU WHC, Western History Collection of the Norlin Library of the University of Colorado; and RMNP, Rocky Mountain National Park Historical Collection. For photographs from published sources, the published citation is given.

Locations of camera points are given by the Township and Range grid system on U.S. Geological Survey topographic maps. References to the locations of objects in a photograph make use of the locational terms given in Figure 11.

Upper left Left background	Upper center Center background	Upper right Right backgound
Left center Left midground	Center Center midground	Right center Right midground
Lower left Left foreground	Lower center Center foreground	Lower right Right foreground

Figure 11. Locational terms used to indicate the positions of objects in the photographs. After Rogers (1982).

The plates are divided into two parts: Part one, Plates 1 through 44, are in Boulder County and adjacent localities; Part two, Plates 45 through 69, are in the Rocky Mountain National Park and adjacent localities.

Plate 1. Rollinsville. *Original*

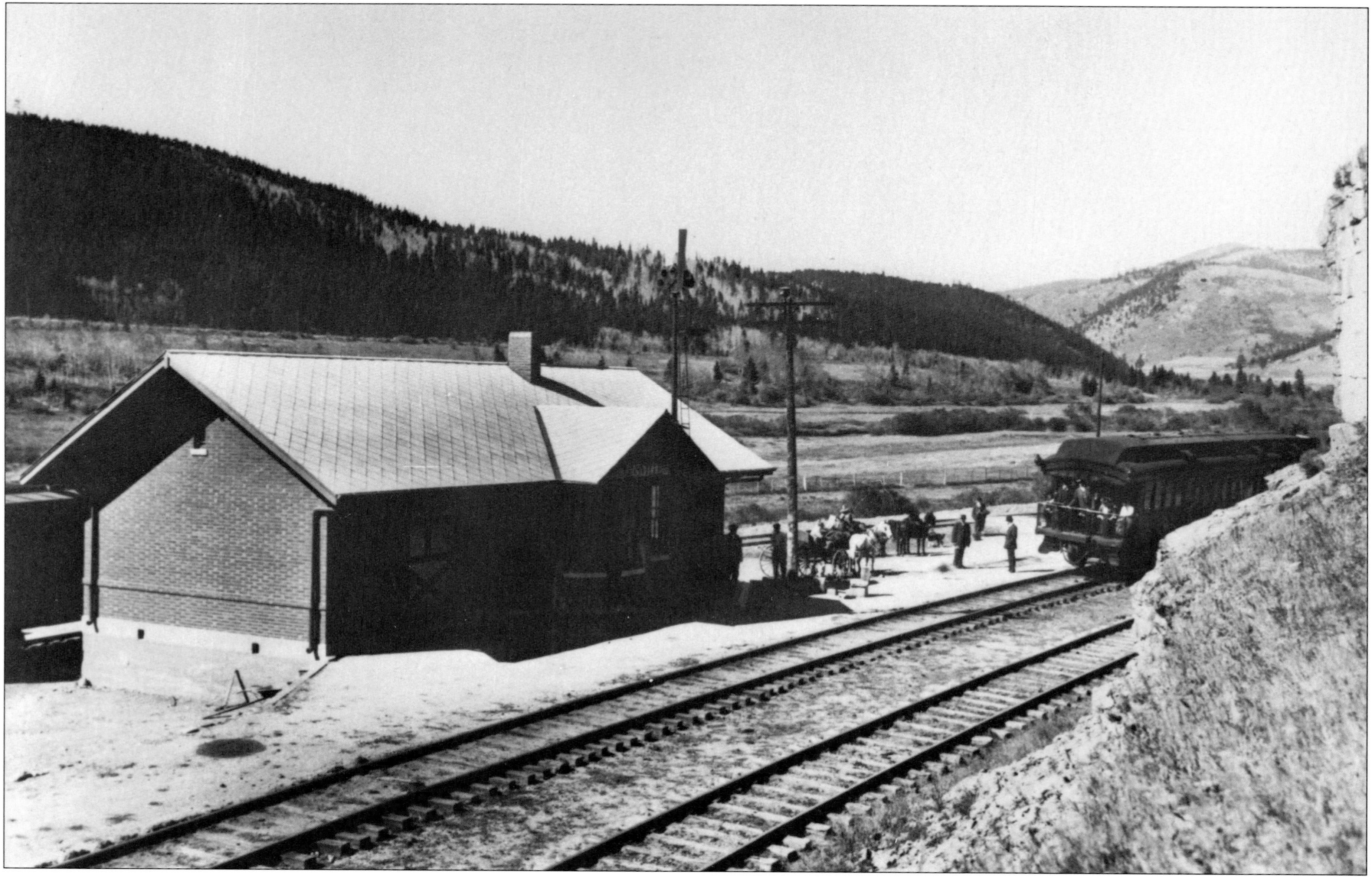

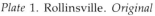

Location: The view is to the southwest taken 1.7 km southeast of the intersection of sections 25 and 35, T1S R73W, at an altitude of 2547 m. It shows the Rollinsville railroad station located along the South Boulder Creek.

Original: c. 1915, L. C. McClure No. 2154, DPL.
Match: 1984, T. T. Veblen and D. C. Lorenz No. G1.
Description: In 1915 the slope to the left and center was covered by a young lodgepole pine and aspen stand. Today, the tree ages and the presence of charcoal on the soil surface indicate that this stand originated following a fire in the early 1850s. Thus, the fire predates the period of mineral exploration

Plate 1. Rollinsville. *Match*

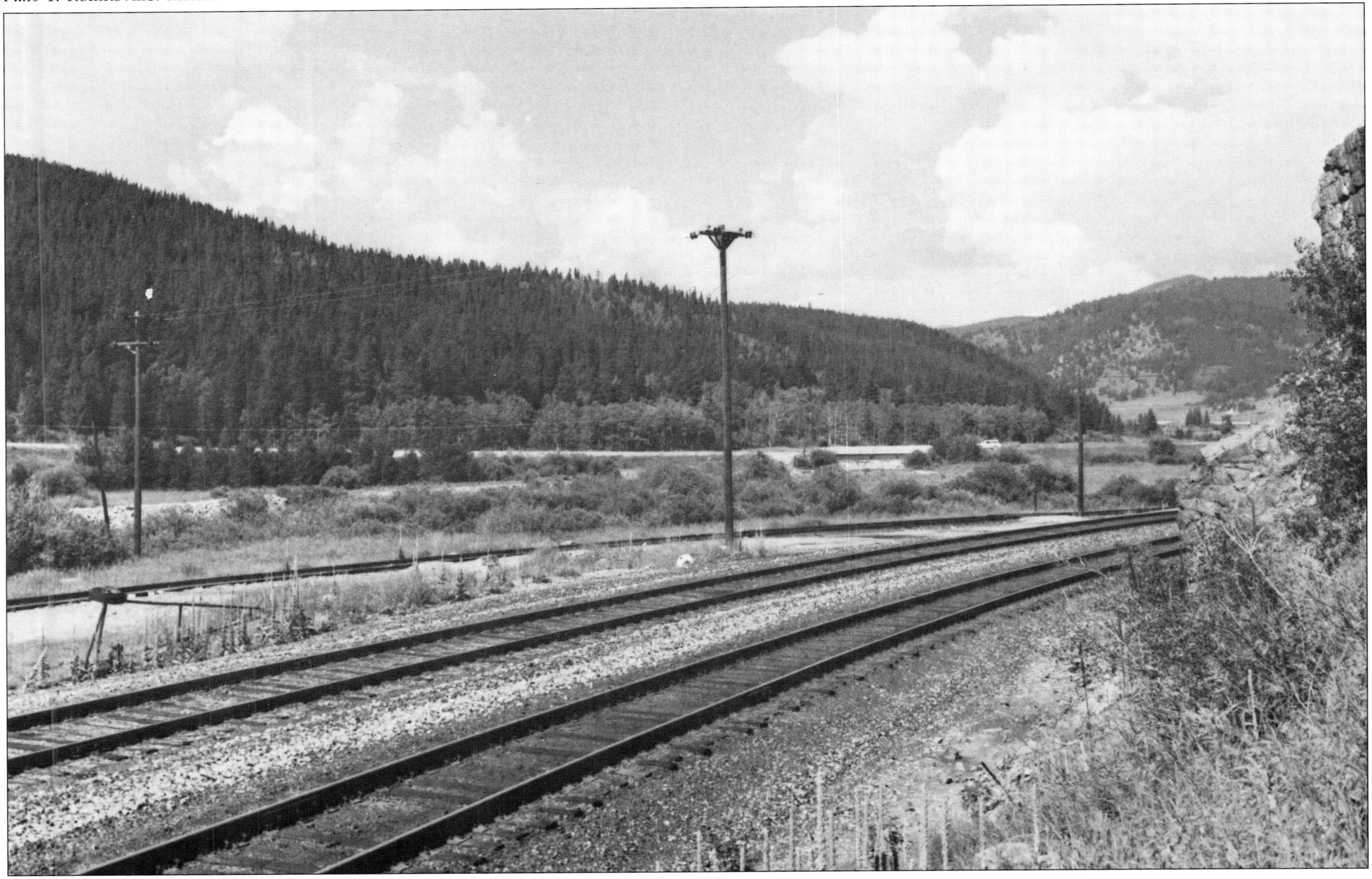

by white prospectors by seven or eight years. The aspen that were prominent in 1915 have been nearly replaced by the lodgepole pine. Three sites along this slope were sampled for forest composition and age structure. Of the 187 trees sampled, 91 percent were lodgepole pine, 5 percent Engelmann spruce, 3 percent Douglas fir, and 1 percent subalpine fir. Lodgepole pine occurred only as older trees while Engelmann spruce and the other species were common as young trees and seedlings. This stand is typical of a postfire stand in which lodgepole pine is being successionally replaced by more shade-tolerant conifers. On the slope in the right background, the burned forest has recovered to a relatively dense cover of conifers.

Plate 2. East side of Rollins Pass. *Original*

Location: The view is to the north taken 0.3 km east of Jenny Lake, at an elevation of 3306 m. The camera point is 4.2 km west-southwest of the intersection of sections 19 and 30, T1S R74W. Yankee Doodle Lake is hidden to the right in the background and Old Wagon Road is barely visible on the far right-hand slope.

Original: c. 1910, L. C. McClure No. 454, DPL.

Match: 1984, T. T. Veblen and D. C. Lorenz No. G5.

Description: This spot near Yankee Doodle Lake was a popular tourist stop on the ride along the

Plate 2. East side of Rollins Pass. *Match*

Moffat Road over the Continental Divide. The dead standing trees just behind the train were killed by a nineteenth-century fire which also burned the slopes in the background. As seen in the modern photograph, the burned slopes have generally recovered to forests dominated by Engelmann spruce and subalpine fir, and by limber pine near timberline on the far right. The Old Wagon Road, constructed in the 1860s, is barely visible in the right background.

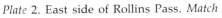

Plate 3. Needle Eye Tunnel and Jenny Lake. *Original*

Location: This is a northeast view of Needle Eye Tunnel and Jenny Lake, taken 0.6 km southwest of Jenny Lake at an elevation of 3209 m. The camera point is 5.3 km west of the intersection of sections 30 and 31, T1S R74W.

Original: c. 1905, L. C. McClure No. 139, DPL.
Match: 1984, T. T. Veblen and D. C. Lorenz No. G6.

Description: The railroad crossed Rollins Pass via the Moffat Road from 1904 to 1929 before being replaced by the 10-km-long Moffat Tunnel. Following the opening of the tunnel, the tracks over

Plate 3. Needle Eye Tunnel and Jenny Lake. *Match*

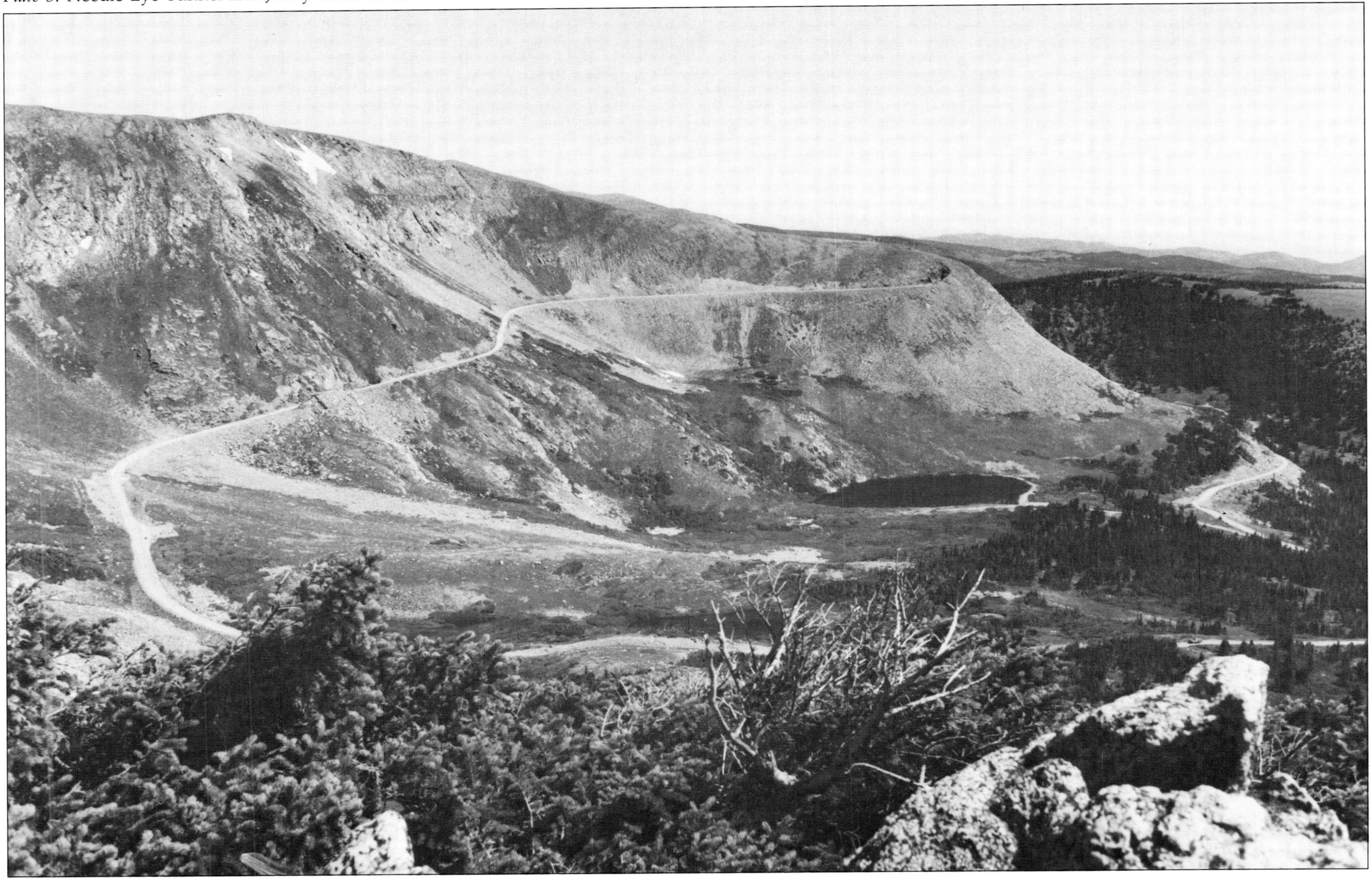

the Moffat Road were removed and the roadbed has since been used by automobiles. Between Jenny Lake and the railway below, the height and density of the Engelmann spruce and subalpine fir krummholz appear to have increased slightly. The upper limit of forest, however, has not changed significantly. In the far right the burn near Old Wagon Road described in Plate 2 is visible, as is the rest stop shown in Plate 2 (right midground).

Plate 4. Near the summit of Rollins Pass. *Original*

Location: This is a northeast view of the Continental Divide from the summit of Rollins Pass taken 6.0 km west of the intersection of sections 19 and 30, T1S R73W, at an elevation of 3559 m.

Original: c. 1905, L. C. McClure No. 124, DPL.
Match: 1984, T. T. Veblen and D. C. Lorenz No. G4.

Description: This is the former location of Corona, a way station and hotel stop on the Moffat Road. Due to the common 7-to-10-m snowdrifts, the tracks in the summit area were covered with

Plate 4. Near the summit of Rollins Pass. *Match*

extensive snowsheds that were dismantled along with the tracks in 1929. In the left center of the 1905 photograph, a path probably created by landslides or snow avalanches extends from the ridge several 100 m down into the forest. In the modern photo, the Engelmann spruce and subalpine fir forest has partially recovered from this disturbance. The gray tones in the left center indicate a large number of dead standing trees in both photographs.

Plate 5. Ptarmigan Point, west side of Rollins Pass. *Original*

Location: This is a view west of the Williams Fork Range taken 8.0 km west of the intersection of sections 30 and 31, T1S R74W, at an elevation of 3489 m.

Original: c. 1905, L. C. McClure No. 123, DPL.
Match: 1984, T. T. Veblen and D. C. Lorenz No. G3.

Description: In the 1905 photograph, the center slopes of Engelmann spruce and subalpine fir had been cleared for the Moffat Road railway. Today, these forests have partially recovered to the same

Plate 5. Ptarmigan Point, west side of Rollins Pass. *Match*

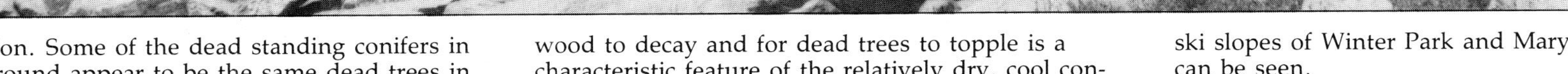

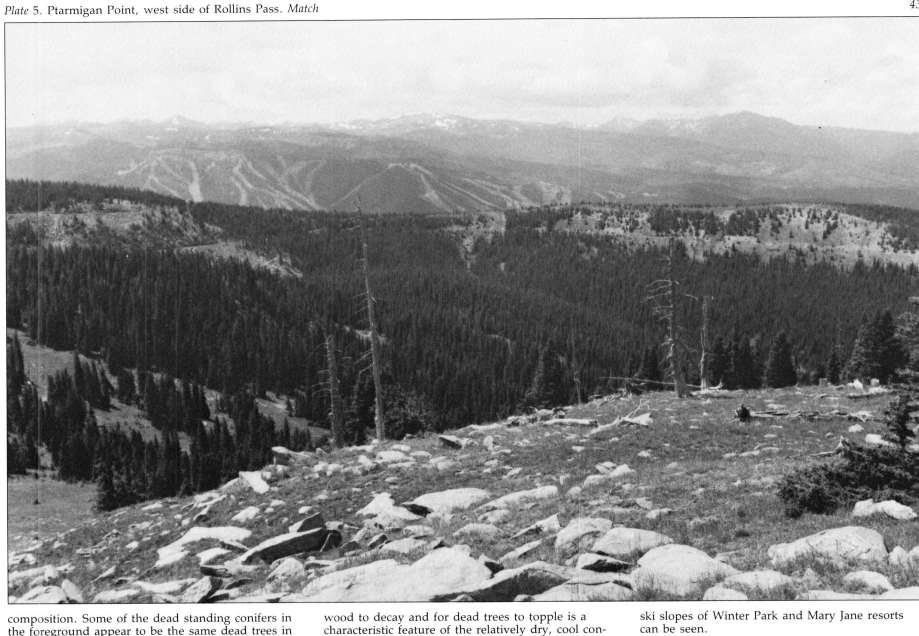

composition. Some of the dead standing conifers in the foreground appear to be the same dead trees in both photographs. The long period required for wood to decay and for dead trees to topple is a characteristic feature of the relatively dry, cool conditions of the Front Range. In the background the ski slopes of Winter Park and Mary Jane resorts can be seen.

Location: This north-northwest view from Chautauqua shows Red Rocks and the Dakota Ridge. The camera point was 0.75 km north-northwest from the intersection of sections 1 and 7, T1S R70W, at an elevation of 1745 m.

Original: c. 1905, L. C. McClure No. 306, DPL.
Match: 1988, T. T. Veblen and D. C. Lorenz No. B54.

Description: This match shows the dramatic increase in abundance of ponderosa pine now typical of the lower montane zone. Slopes in the upper left of the original photograph that were open woodlands

Plate 6. Chautauqua, Boulder. *Match*

in the early twentieth century are now dense stands of ponderosa pine. Slopes that formerly were pure grasslands (upper right) today support scattered ponderosa pine and Rocky Mountain juniper. Planted trees have created a substantial tree cover in the urban area of Boulder. Chautauqua was established in 1897 by a group of Texans seeking respite from the heat and dust of Texas. Initially a tent camp, the construction of a tabernacle, which became the present auditorium began in 1898.

Plate 7. Red Rocks, Boulder. *Original*

Location: This is a view north of Red Rocks at the mouth of Boulder Canyon. It was taken 0.3 km southeast from the intersection of sections 35 and 25, T1N R71W, at an elevation of 1647 m.
Original: c. 1905, J. P. Sturtevant No. 2258, BHS.
Match: 1985, T. T. Veblen and D. C. Lorenz No. B34.
Description: The Aikins expedition, the first white

Plate 7. Red Rocks, Boulder. *Match*

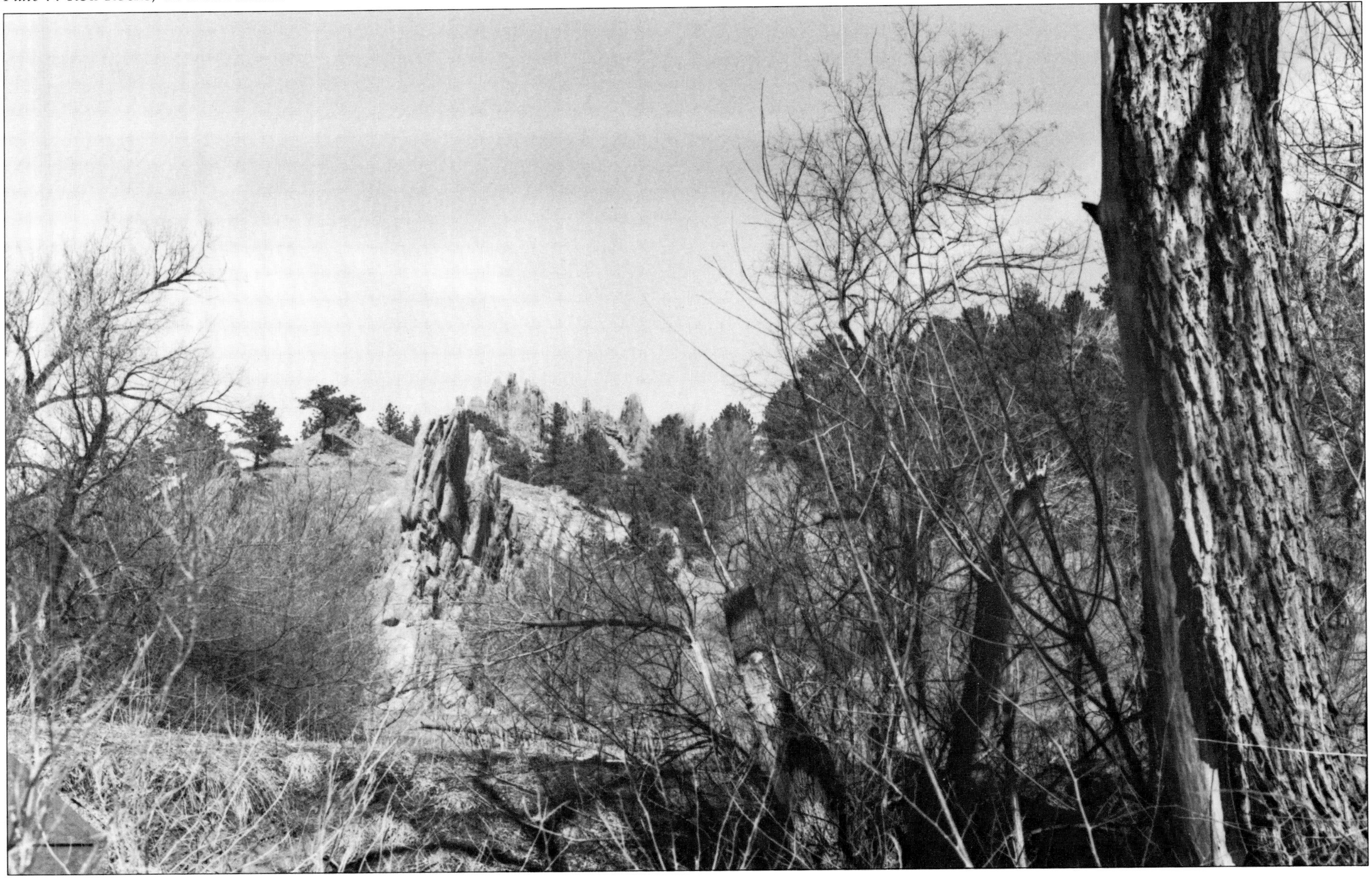

settlers of Boulder, are believed to have initially camped at the foot of these sandstone outcrops. Recent road construction prevented an exact match of the old and modern photographs. In the foreground there has been a marked increase in the abundance of narrow-leaf cottonwood along Boulder Creek. Ponderosa pine has also increased on the sandstone outcrops and in the surrounding grassland.

Plate 8. The Dakota Ridge, Boulder. *Original*

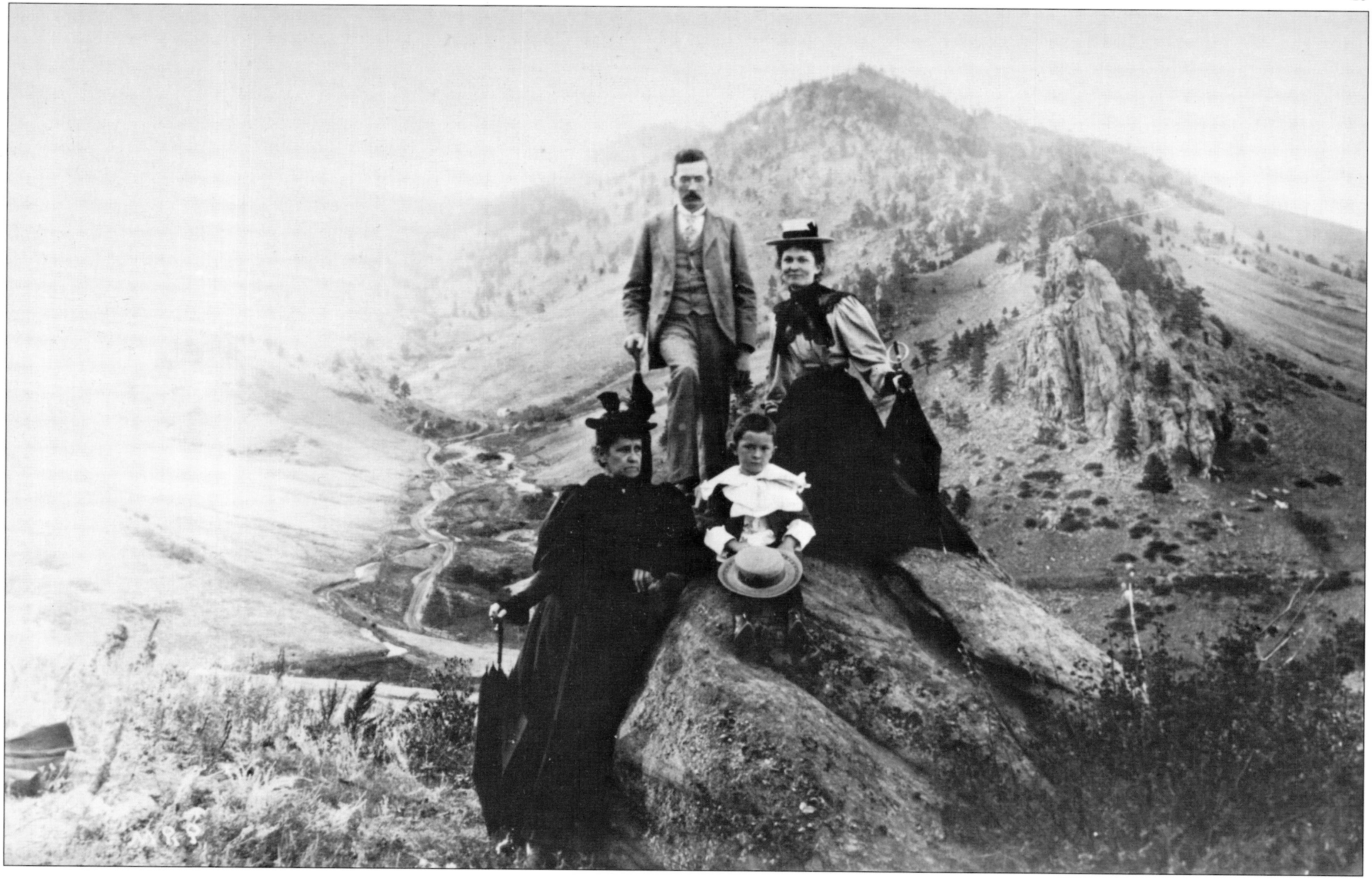

Location: This is a view north of the Dakota Ridge from the north slope of Red Rocks. The photo was taken 0.4 km northeast of the intersection of sections 35 and 25, T1N R71W, at an elevation of 1647 m.

Original: c. 1900, J. P. Sturtevant No. 1241, BHS.

Match: 1985, T. T. Veblen and D. C. Lorenz No. B35.

Description: This match shows a distinct increase in

Plate 8. The Dakota Ridge, Boulder. *Match*

the abundance of ponderosa pine, so areas that formerly were grassland are now woodlands. The increase in tree density has been greatest along the bases of rocky outcrops where the soils are coarse textured and more favorable to tree seedlings than to grass.

Plate 9. Flagstaff Mountain, view west. *Original*

Location: This is a view west of Flagstaff Mountain from the base of Red Rocks. The photograph was taken 0.3 km southeast of the intersection of sections 35 and 25, T1N R71W, at an elevation of 1647 m.

Original: c. 1900, J. P. Sturtevant No. 2203, BHS.

Match: 1985, T. T. Veblen and D. C. Lorenz No. B39.

Description: The slopes of Flagstaff Mountain have

Plate 9. Flagstaff Mountain, view west. *Match*

been filled in by ponderosa pine, particularly on the northwest-facing slopes and along ridgetops. In the foreground, the cottonwoods along Boulder Creek are much larger. Road construction and embankment of the Boulder Creek prevented an exact match.

Plate 10. Eldorado Springs. *Original*

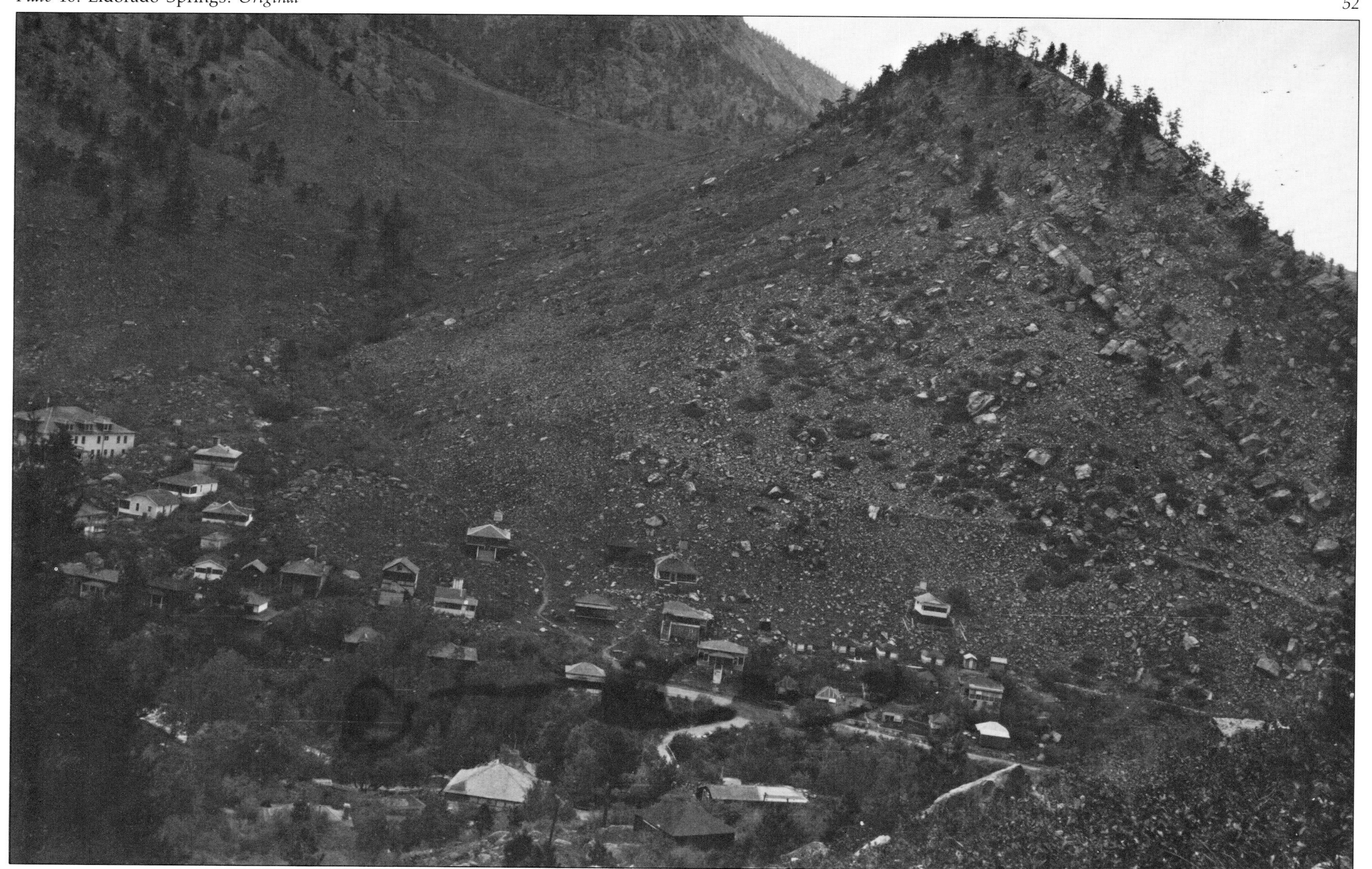

Location: This is a view north of Eldorado Springs, located 6.5 km south of Boulder. The photo was taken 0.05 km east of the intersection of sections 30 and 31, T1S R70W, at an elevation of 1854 m. *Original:* 1921, W. T. Lee No. 1784, USGS.

Match: 1985, T. T. Veblen and D. C. Lorenz No. B44.

Plate 10. Eldorado Springs. *Match*

Description: In the original photograph ponderosa pine occurs in sparse stands in the ravines (center of photo) and along ridgetops (upper right). The present photograph shows a substantial increase in stand density in these areas and an expansion of ponderosa pine and Rocky Mountain juniper into grasslands.

Plate 11. Red Hill Valley, Boulder. *Original*

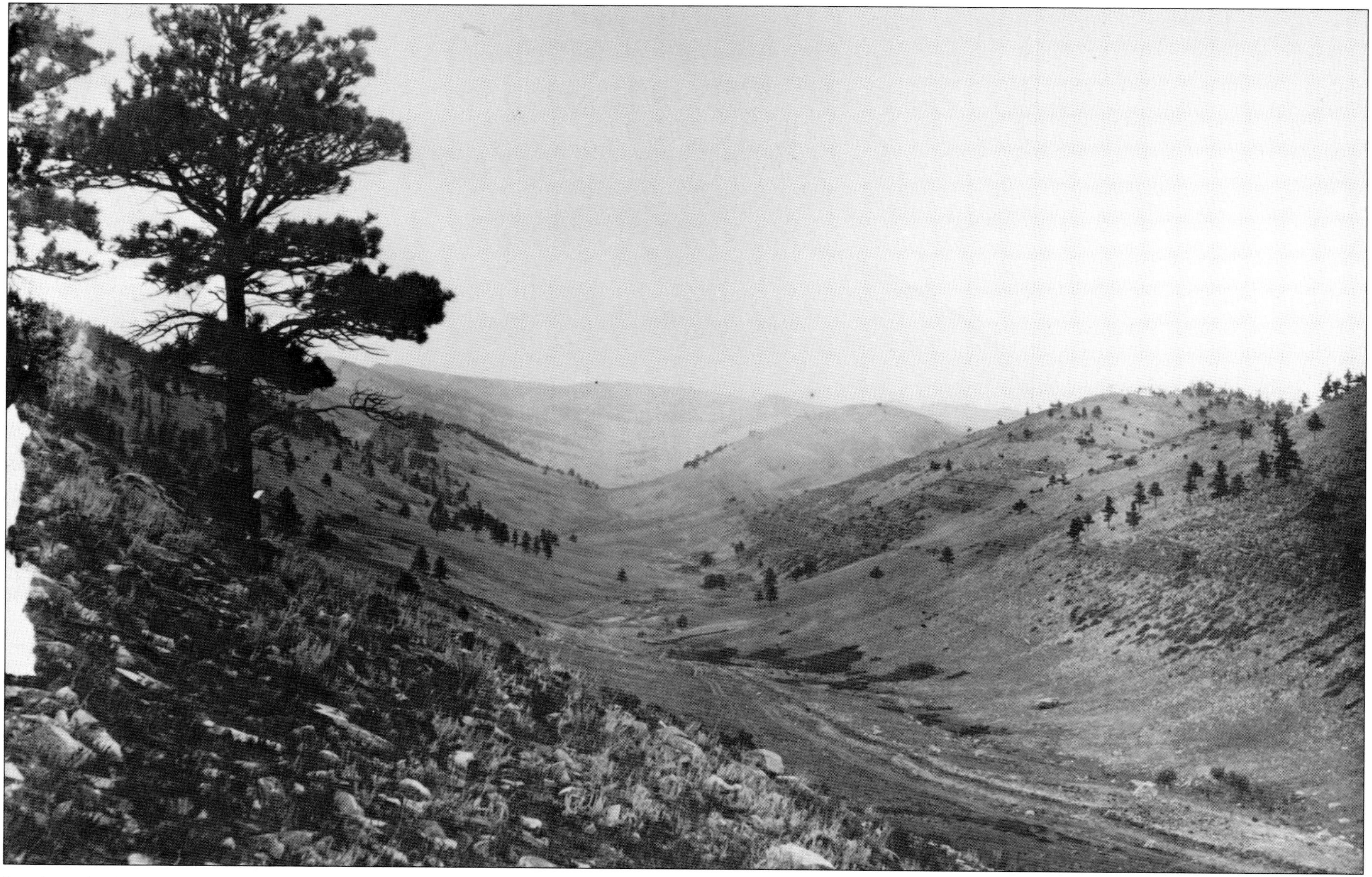

Location: This is a view north of Red Hill Valley, northwest of Boulder. The camera point was 0.2 km west of the intersection of sections 2 and 36, T1N R71W, at an elevation of 1891 m.

Original: 1899, Paddock Collection, BHS.
Match: 1984, T. T. Veblen and D. C. Lorenz No. B30.

Description: The construction of a house at the original camera point prevented an exact match. The valley has generally experienced a marked

Plate 11. Red Hill Valley, Boulder. *Match*

increase in the abundance of ponderosa pine. Sparse groups of ponderosa pine in the original photograph have become dense patches. The distant slope in the upper center has changed from open woodland to a dense forest. In the mid-ground and foreground, the lack of stumps or logs in the original photograph indicates that this area was not forested during the late nineteenth century.

Plate 12. Mouth of Boulder Creek. *Original*

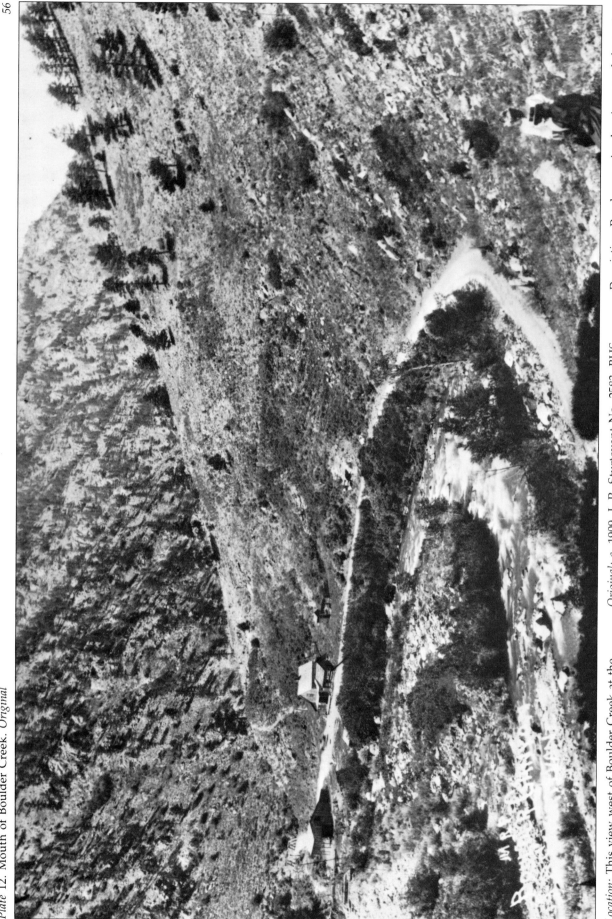

Location: This view west of Boulder Creek at the head of Boulder Canyon was taken 0.05 km east of the intersection of sections 35 and 25, T1N R71W, at an elevation of 1696 m.

Original: c. 1909, J. P. Sturtevant No. 2583, BHS.
Match: 1985, T. T. Veblen and D. C. Lorenz No. B40.

Description: Ponderosa pine has increased dramatically on both the north- and south-facing slopes. Originally, the south-facing slopes were nearly treeless but now support ponderosa pine

Plate 12. Mouth of Boulder Creek. *Match*

woodland. The north-facing slopes are presently covered by dense ponderosa pine stands, with Douglas fir common in moister ravines. The house above the road in the original photograph was the toll house for the Boulder Canyon Road.

Plate 13. McAllister Sawmill. *Original*

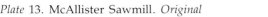

Location: This is a view north of the sawmill, located in Boulder Canyon, 13 km west of Boulder. The photo was taken 0.4 km north of the intersection of sections 10 and 2, T1S R72W, at an elevation of 2306 m.

Original: 1897, J. P. Sturtevant No. 139, BHS.

Match: 1985, T. T. Veblen and D. C. Lorenz No. B31.

Description: Timber on the slopes to the left center and in the background was cut to supply the McAllister Sawmill. Cutting and burning, as indicated by spars, resulted in a fairly open stand of

mainly Douglas fir on these northeasterly slopes at the time of the original photograph. A stand in the left center of the photograph was sampled for age structure and composition. The stand consisted mainly of Douglas fir less than 90 years old. Remnant Douglas fir up to nearly 200 years old were also common. Small numbers of Ponderosa pine, both young individuals and a few remnant trees, were also present. Comparison of the identities of stumps cut early in this century with the present stand suggests that the species composition of the site has not changed significantly.

Plate 14. Sunshine. *Original*

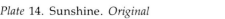

Location: This is a view to the northwest of Sunshine at the head of Sunshine Canyon. The photo was taken 0.55 km northwest of the intersection of sections 17 and 9, T1N R71W, at an altitude of 2196 m.
Original: 1905, J. P. Sturtevant No. 1739, BHS.
Match: 1985, T. T. Veblen and D. C. Lorenz No. B33.
Description: The town of Sunshine, which sup-

ported a population of 1500 in 1876, was an important production center of tellurium. The principal mine was located in the upper right of the photograph. Most of the open meadow in the original photograph is now covered with woodlands and forest of ponderosa.

Plate 15. Salina. *Original*

Location: This is a view east of Salina, located approximately 7.0 km from the junction of Boulder Canyon and Four Mile Canyon. The photo was taken 0.77 km southeast of the intersection of sections 13 and 7, T1N R71W, at an altitude of 2111 m.

Original: c. 1900, J. P. Sturtevant No. 1664, BHS.
Match: 1984, T. T. Veblen and D. C. Lorenz No. B19.

Plate 15. Salina. *Match*

Description: Recent construction prevented an exact match. Ponderosa pine has increased in density on the center midground and background slopes. Willows and cottonwoods are now conspicuous along Four Mile Creek to the lower right. This area was severely affected by the 1894 flood that destroyed much of downtown Boulder.

Plate 16. Ward line near Sunset. *Original*

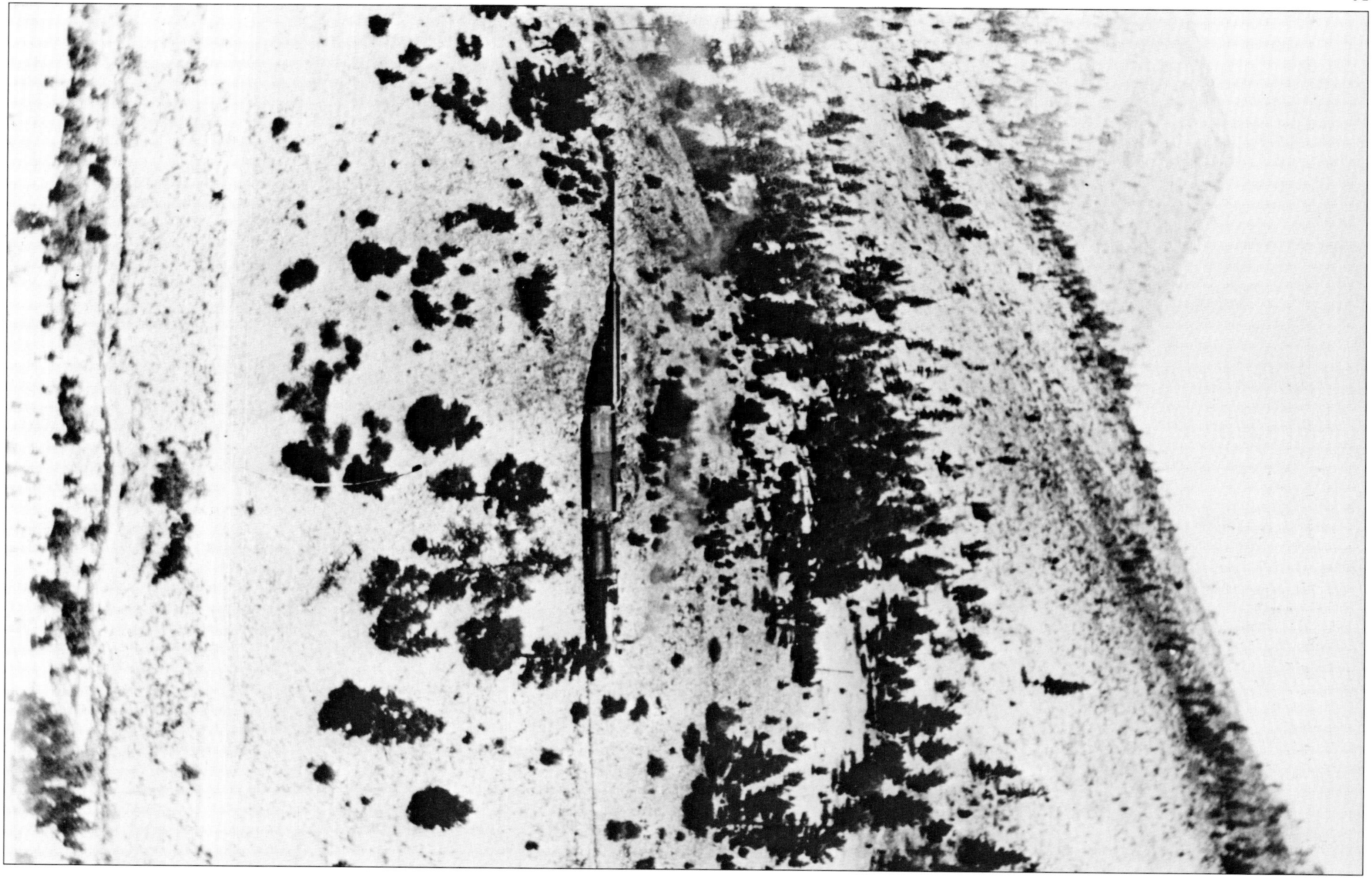

Location: This is a view north from the Eldora line to the Ward line, with Sunset located 1.0 km southeast of the photo. The photo was taken 0.55 km to the north of the intersection of sections 21 and 24, T1N R72W, at an elevation of 2379 m.
Original: c. 1905, H. Sherman, Crossen (1978).
Match: 1984, T. T. Veblen and D. C. Lorenz No. B27.

Description: This south-facing slope had been burned and logged in the late nineteenth century. Timber was used both for the construction of Sunset and for the many mines in this area. The

Plate 16. Ward line near Sunset. *Match*

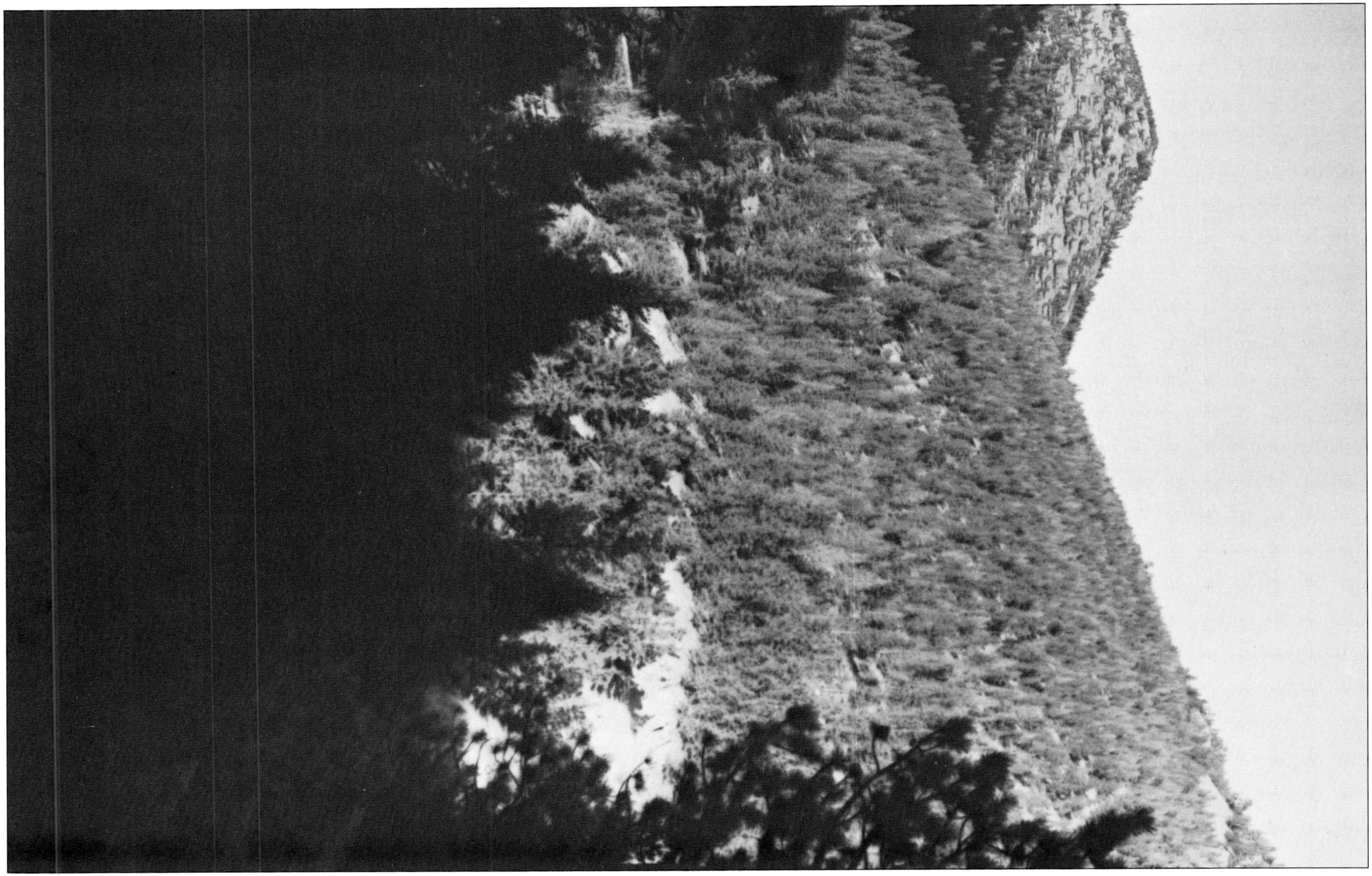

match shows good recovery of the disturbed forest to a dense stand. Of the trees sampled in a plot in the center of this photograph, 65 percent were ponderosa pine, 24 percent Douglas fir, and the remainder were Rocky Mountain juniper or limber pine. Nearly all trees sampled were less than 90 years old. The gray crowns of many trees reflect defoliation of Douglas fir by spruce budworm during the early 1980s.

Location: A view northeast of the town of Sunset taken 0.5 km southeast along the Eldora line. The photo was taken 0.5 km southwest of the intersection of sections 16 and 20, T1N R72W, at an elevation of 2356 m.
Original: 1898, J. P. Sturtevant, Crossen (1978).

Match: 1984, T. T. Veblen and D. C. Lorenz No. B7.
Description: The recent construction of a house

Plate 17. Sunset. *Match*

prevented an exact match. The slope in the upper right had been recently burned at the time of the 1898 photograph, as indicated by the abundant spars. It has recovered to a Douglas fir–dominated stand which in the early 1980s was severely attacked by spruce budworm. On the more southerly facing slope, in the midground and foreground, ponderosa pine has increased its coverage.

Plate 18. Giant's Ladder. *Original*

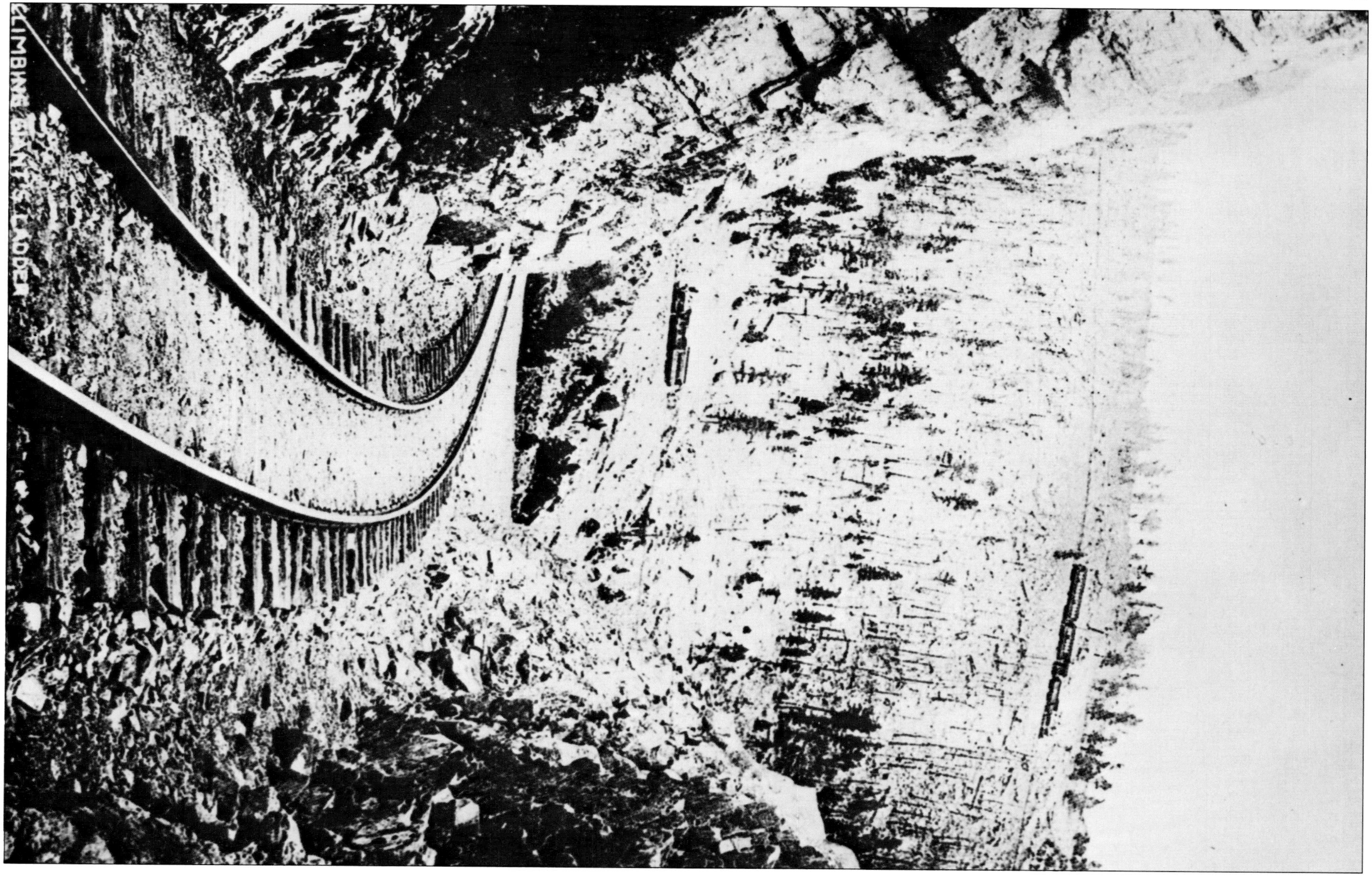

Location: This is a view north, 1.0 km northeast of Sunset along the Ward line. The photo was taken 0.4 km southeast of the intersection of sections 16 and 20, T1N R72W, at an elevation of 2471 m.
Original: 1905, Paddock Collection, Crossen (1978).

Match: 1982, N. R. Johnson No. B59.
Description: This is the beginning of the Switzerland Trail's climb from Sunset to Frankeberger

Plate 18. Giant's Ladder. *Match*

Point. Most of the slope in the center had been recently burned at the time of the original photograph. Today, it has recovered to a ponderosa pine-dominated stand. Numerous cut stumps also indicate abundant logging of this slope during the late nineteenth and early twentieth centuries.

Plate 19. Frankeberger Point. *Original*

Location: This is a view south-southwest from Frankeberger Point, 2.5 km northeast of Sunset along the Ward line. The camera point was 0.4 km northeast of the intersection of sections 15 and 21, T1N R72W, at an elevation of 2579 m.

Original: 1905, J. P. Sturtevant, Crossen (1978).
Match: 1984, H. Malde No. B12.
Description: The slopes in the background had been extensively logged and burned in the late nineteenth century. The large number of nineteenth-century mines in Four Mile Canyon produced a heavy demand for timber for mine construction and fuel. Comparison with the modern photograph shows a substantial recovery of ponderosa pine

Plate 19. Frankeberger Point. *Match*

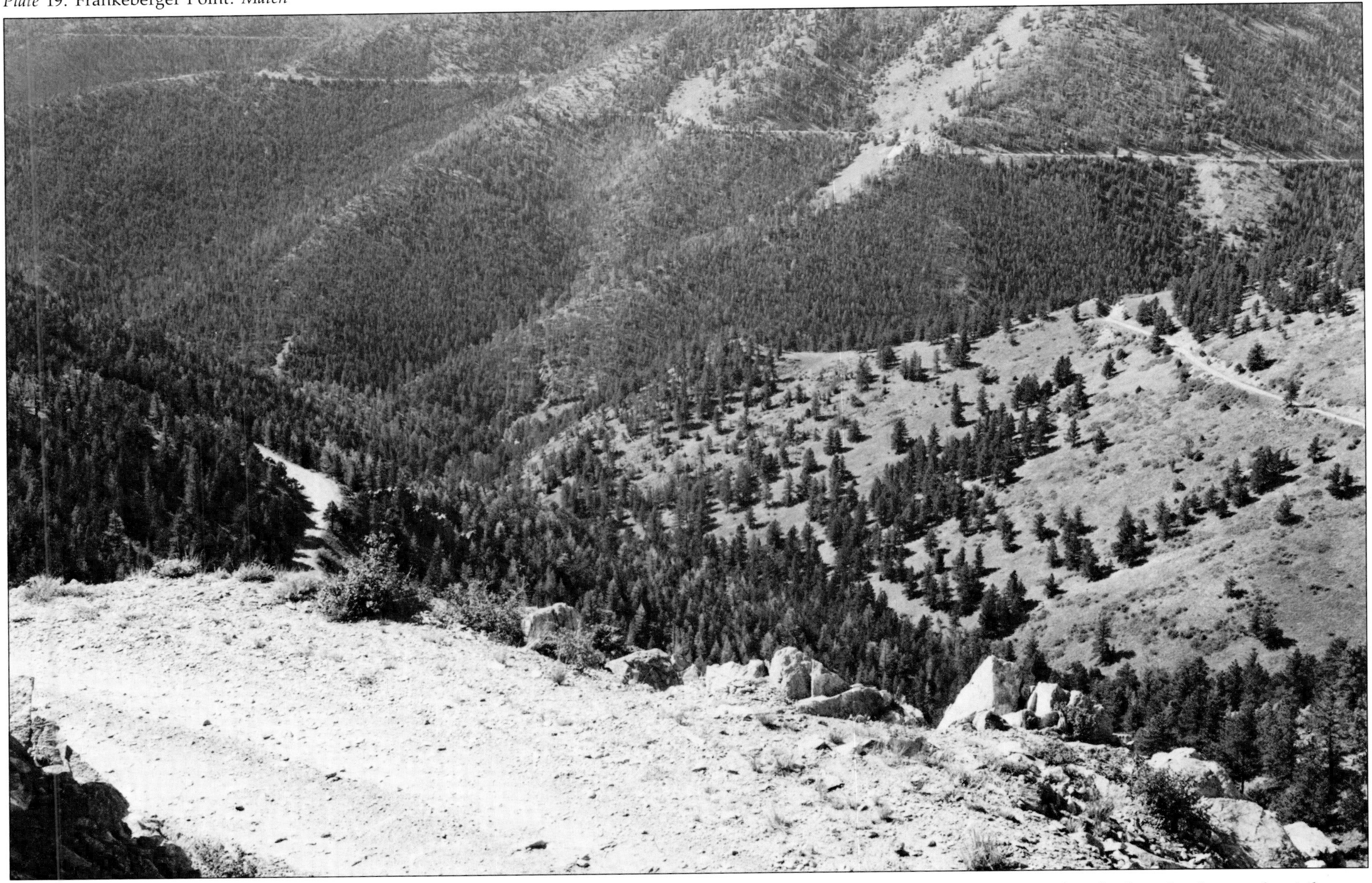

and Douglas fir stands on this northerly slope. The gray tones in the modern photo are Douglas firs recently defoliated by spruce budworm. Note the grassland just below the railroad notch on the slope in the right center in the original photograph. In the modern photograph this slope has been invaded by ponderosa pine. A sample of this stand indicates that ponderosa pine accounts for over 90 percent of the trees and Douglas fir and Rocky Mountain juniper for the remainder. Tree ages indicate that the invasion began about the turn of the century. Similar invasions of montane grasslands by ponderosa pine are common throughout the Front Range.

Plate 20. Ward line in Four Mile Canyon. *Original*

Location: This curve on the Switzerland Trail is located 1.5 km northeast of Sunset. The photo was taken 0.2 km northeast of the intersection of sections 15 and 21, T1N R72W, at an elevation of 2486 m.
Original: c. 1900, J. P. Sturtevant, Crossen (1978).

Match: 1984, H. Malde No. B10.
Description: The slope in the background was partially burned and logged in the late nineteenth

Plate 20. Ward line in Four Mile Canyon. *Match*

century. It has now recovered mainly to ponderosa pine. A sample taken in the right center on the southerly exposure indicates a present composition of 90 percent ponderosa pine and 10 percent Douglas fir. These are largely postfire populations which began establishing in the 1890s, although today there are still scattered remnant trees well over 100 years old.

Plate 21. Mt. Alto, view southeast. *Original*

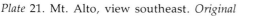

Location: This is a view of the Oxbow curve from Mt. Alto, 7.2 km northeast of Sunset along the Ward line. The camera point was 0.6 km northeast of the intersection of sections 15 and 21, T1N R72W, at an elevation of 2532 m.

Original: 1905, J. P. Sturtevant, CU WHC.
Match: 1982, N. R. Johnson No. B18.
Description: A cut stump and felled trees in the foreground of the original photograph indicate recent cutting at the turn of the century. The sparse stand of conifers on the ridge in the left background also indicates recent logging. Their growth form indicates that they grew in a relatively closed stand (i.e., most lack lower limbs). A sample along this ridge indicates that the present

Plate 21. Mt. Alto, view southeast. *Match*

stand consists of ponderosa pine (88 percent) and Douglas fir (12 percent). Most trees established about the time the photograph was taken in 1905. Numerous remnant Douglas fir over 150 years old and cut stumps of this species suggest that prior to disturbance at the end of the last century, this species was more common here than it is today. Thus, it is likely that this site was initially codominated by relatively old Douglas fir and ponderosa pine and that recovery following logging shifted the dominance towards ponderosa pine.

Plate 22. Oxbow curve. *Original*

Location: This bird's eye north-northeast view of the Oxbow curve was taken 0.25 km west of the intersection of sections 14 and 22, T1N R72W, at an elevation of 2618 m.

Original: 1900, Kindig Collection, Crossen (1978).
Match: 1982, N. R. Johnson No. B61.
Description: In the background of the original photograph, the slopes are relatively open, but today they have been filled in primarily by ponderosa pine. The lack of spars implies that these

Plate 22. Oxbow curve. *Match*

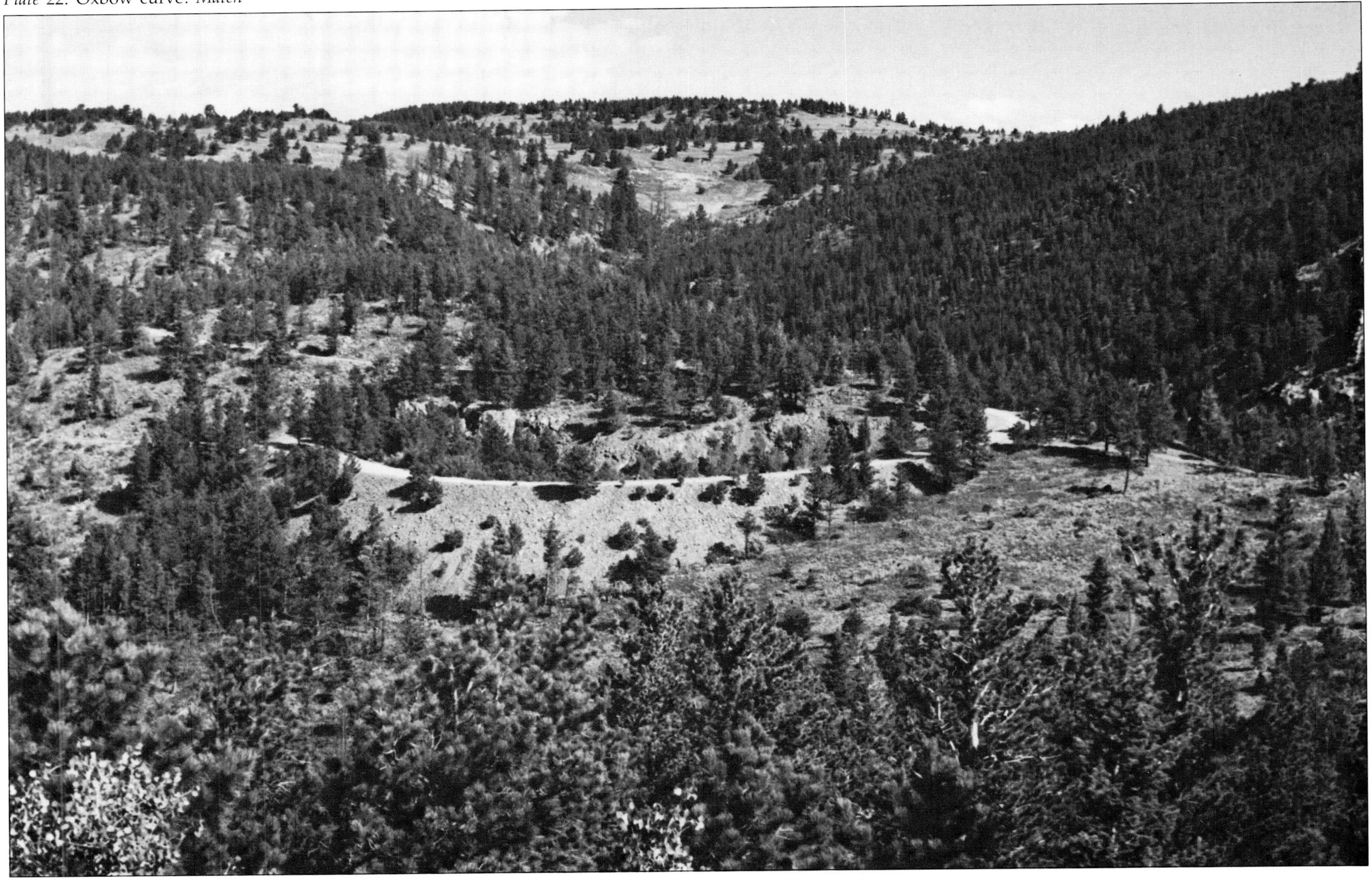

distant slopes were largely treeless prior to Euro-American settlement. In contrast, the southeasterly slope in the right center had been logged and burned in the nineteenth century but today has recovered to a dense stand of ponderosa pine and Douglas fir. The ridge adjacent to the track and to the left of center is the same ridge described for Plate 21.

Plate 23. Oxbow depression. *Original*

Location: The Oxbow depression is located 3.0 km southeast of Gold Hill Station. The photo was taken 0.4 km northwest of the intersection of sections 14 and 22, T1N R72W, at an elevation of 2538 m.
Original: 1900, J. P. Sturtevant, Crossen (1978).

Match: 1982, N. R. Johnson No. B13.
Description: This is the same curve visible in Plate 22. This rocky depression was created by the

Plate 23. Oxbow depression. *Match*

railroad construction in the late 1890s. The match shows dramatic succession from bare ground to a stand of ponderosa pine and Douglas fir. A sample taken here indicates that ponderosa pine accounts for 84 percent of the trees and ranges in age from 25 to 85 years. Douglas fir, which accounts for 16 percent of the trees, ranges in age from 25 to 65 years. Thus, colonization of the bare artificial sites was initially by ponderosa pine followed some 20 years later by Douglas fir.

Plate 24. Oxbow curve, view southwest. *Original*

Location: This view southwest from the Oxbow curve was taken 0.4 km northwest of the intersection of sections 14 and 22, T1N R72W, at an elevation of 2532 m.

Original: 1900, J. P. Sturtevant, Crossen (1978).
Match: 1984, H. Malde No. B62.

Plate 24. Oxbow curve, view southwest. *Match*

Description: This view depicts the same stand in the interior of the curve described for Plate 23. The slopes in the center show signs of extensive burning in the late nineteenth century but have recovered to dense stands of Douglas fir and ponderosa pine, especially on the northerly slopes.

Plate 25. Gold Hill Station. *Original*

Location: This view west from Gold Hill station was taken 0.6 km southwest from the intersection of sections 10 and 16, T1N R72W, at an elevation of 2631 m.

Original: 1899, J. P. Sturtevant, CU WHC.
Match: 1984, T. T. Veblen and D. C. Lorenz No. B56.

Description: This was the Switzerland Trail station for the important mining town of Gold Hill a few kilometers to the east. The numerous cut stumps

Plate 25. Gold Hill Station. *Match*

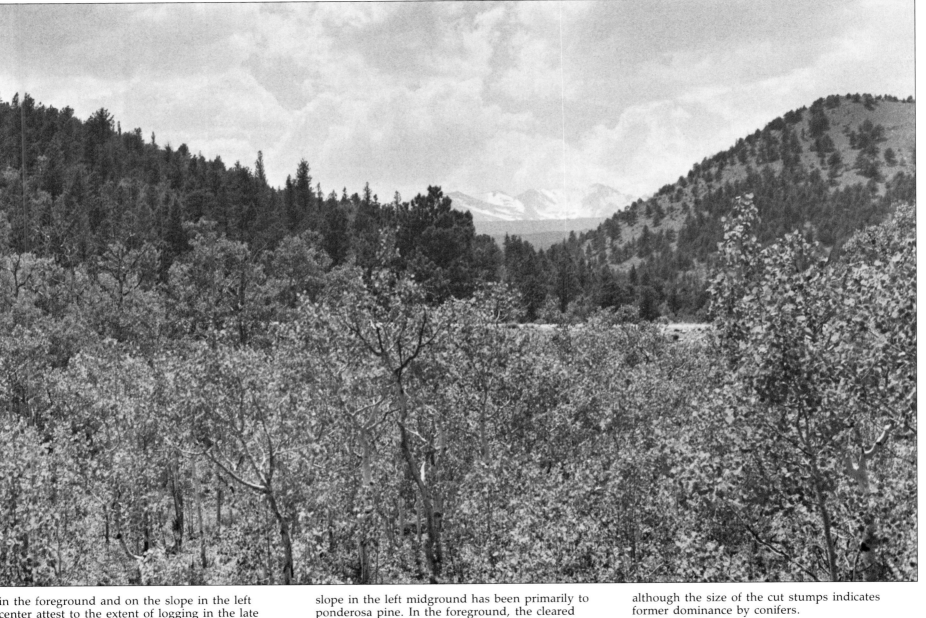

in the foreground and on the slope in the left center attest to the extent of logging in the late nineteenth century. The recovery on the northerly slope in the left midground has been primarily to ponderosa pine. In the foreground, the cleared depression has recovered to quaking aspen, although the size of the cut stumps indicates former dominance by conifers.

Plate 26. Camp Frances, view west. *Original*

Location: This view west shows a portion of the town of Camp Frances on the Switzerland Trail. The camera point was 0.95 km southeast of the intersection of sections 1 and 11, T1N R73W, at an elevation of 2791 m.

Original: 1898, L. Meile, Crossen (1978).
Match: 1984, H. Malde No. B14.
Description: This settlement had a population of nearly 200 people at the turn of the century and served the several nearby silver mines. The foreground was extensively logged, as was the center midground slope. The upper slopes had been extensively burned prior to the time of the photograph in 1898. Four stands were sampled on the center midground and the upper slopes in the

Plate 26. Camp Frances, view west. *Match*

center background. Today the center midground slope is dominated by lodgepole pine mostly 55 to 95 years old, with small numbers of ponderosa pine and limber pine also present. Numerous cut stumps are still present in these stands. The contemporary forest on the upper slope is more heterogeneous in composition and slightly older. Subalpine fir accounted for 44 percent of the trees, lodgepole pine for 35 percent, Engelmann spruce for 11 percent, and limber pine for 10 percent. These postfire stands began establishing about 1880. Scattered remnant trees over 150 years old also occur in the postfire stand. The stand in the depression in the foreground is dominated by limber pine and aspen, as described in Plate 27.

Plate 27. Camp Frances, view east-southeast. *Original*

Location: This photo was taken 0.8 km southeast of the intersection of sections 1 and 11, T1N R73W, at an elevation of 2806 m.

Original: 1898, J. P. Sturtevant, Crossen (1978).
Match: 1984, H. Malde No. B15.

Description: This is a view of the aspen and limber pine stand visible in the foreground of Plate 26. The site had been clearcut by the end of the

Plate 27. Camp Frances, view east-southeast. *Match*

nineteenth century. A sample taken here indicates that aspen and limber pine are the most common trees, with a few Douglas fir also present. Tree establishment began about 1900. The abundance of conifer stumps indicates that logging resulted in a shift in species composition towards dominance by aspen, which root-suckers vigorously after logging.

Plate 28. Ward, view southeast. *Original*

Location: This view of Ward was taken 0.2 km northwest of the intersection of sections 6 and 12, T1N R73W, at an elevation of 2837 m.

Original: 1900, unknown photographer, BHS.

Match: 1982, N. R. Johnson No. B65.

Description: This view of the major mining town of

Plate 28. Ward, view southeast. *Match*

Ward, established in 1860, shows it at its heyday just before its nearly complete destruction by fire in 1900. In the late nineteenth century, the slopes in the background had been extensively logged. Today, these slopes are forested largely by lodgepole pine.

Plate 29. Ward, view north. *Original*

Location: This view north from Ward was taken 0.25 km northwest from the intersection of sections 6 and 12, T1N R73W, at an elevation of 2806 m.

Original: 1870, J. P. Sturtevant, CHS.
Match: 1985, T. T. Veblen and D. C. Lorenz No. B53.

Description: This early view of Ward shows that the forest adjacent to the town was still largely intact despite the abundant piles of logs present in the

Plate 29. Ward, view north. *Match*

town. Comparison with Plate 28 indicates the magnitude of the deforestation resulting from fuel and construction needs over the following 30 years. This southerly slope, also visible in the left center of Plate 28, was mostly cut following the 1870 photograph and has since recovered to ponderosa pine on the lower slope and lodgepole pine with scattered Engelmann spruce and subalpine fir on the upper slope.

Plate 30. Northwest slope of Sugarloaf Mountain. *Original*

Location: Sugarloaf Mountain is located 2.5 km northwest of the town of Sugarloaf along the Eldora line of the Switzerland Trail. The photo was taken 0.5 km north of the intersection of sections 22 and 28, T1N R72W, at an elevation of 2446 m.

Original: c. 1905, J. P. Sturtevant No. 2418, BHS.
Match: 1984, T. T. Veblen and D. C. Lorenz No. B26.
Description: This section of the Eldora line of the Switzerland Trail in Four Mile Canyon was the hub of extensive gold mining from the 1860s to about 1905. Most of the forest had been burned and cut in the late nineteenth century. Three stands were sampled and indicate that the forest

Plate 30. Northwest slope of Sugarloaf Mountain. *Match*

has recovered largely to Douglas fir with smaller numbers of ponderosa pine and limber pine. Remnant trees over 200 years old and cut stumps of Douglas fir, suggest that the burning did not result in a major successional shift in species composition. Visible in the original photograph is a patch of trees that survived the fires, located on a particularly rocky site where a lower stand density may have prevented a crown fire from developing. In 1984, ponderosa pine and limber pine were more common in this remnant patch than they are elsewhere on the Douglas fir–dominated slope. The slope in the left background burned again in July 1989.

Plate 31. Lee Mine. *Original*

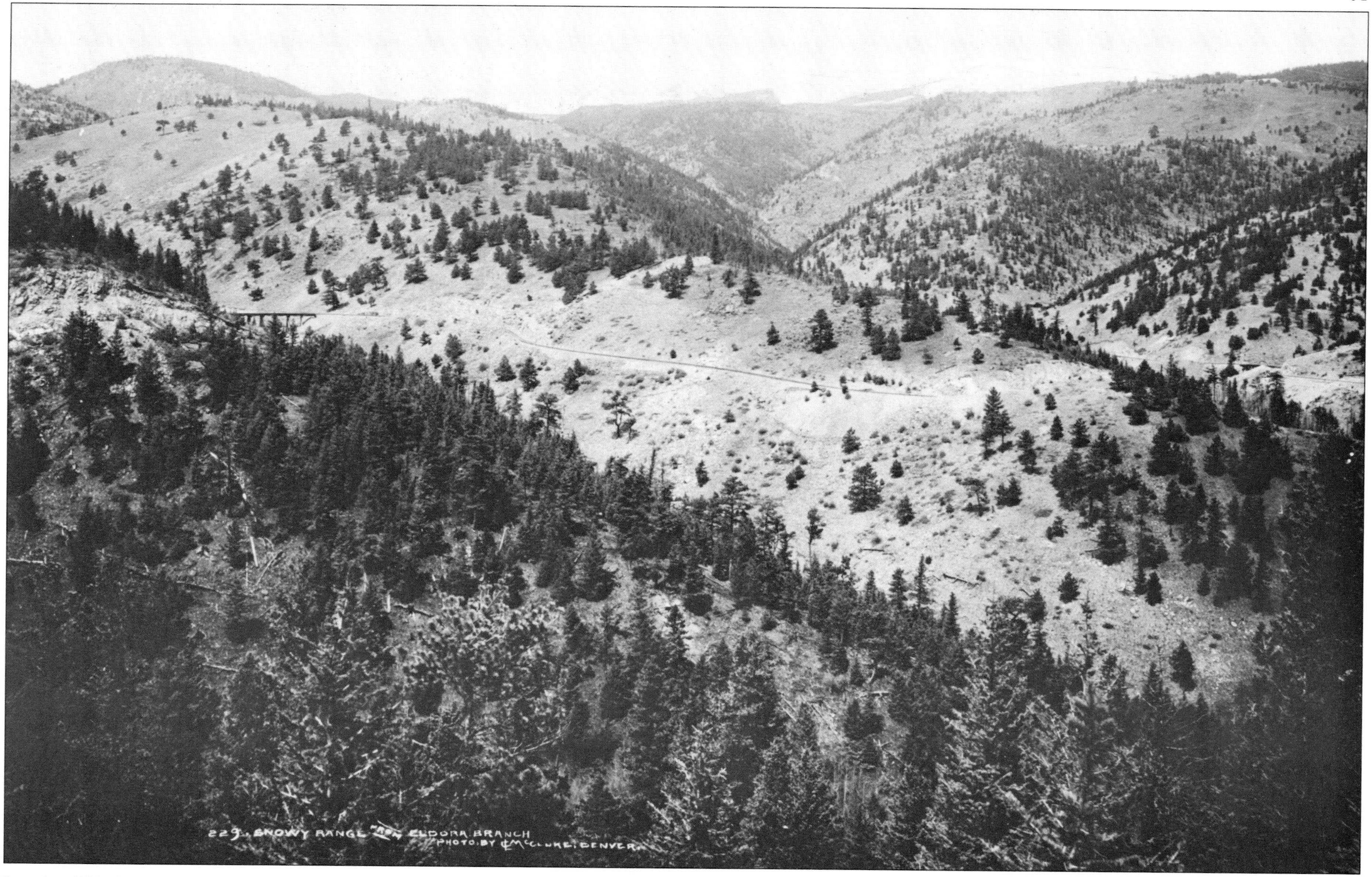

Location: This is a view west along the Eldora line, 4.0 km northwest of Sugarloaf. The camera point was 0.8 km northeast of the intersection of sections 21 and 29, T1N R72W, at an altitude of 2501 m.
Original: c. 1905, L. C. McClure No. 229, DPL.

Match: 1985, T. T. Veblen and D. C. Lorenz No. B51.
Description: The slopes in the center background of

Plate 31. Lee Mine. *Match*

the original photograph were grassland or open woodlands of ponderosa pine. The match shows that the grasslands have been invaded by ponderosa pine and that former woodlands are now dense stands of ponderosa pine and Douglas fir.

Plate 32. Lee Mine. *Original*

Location: This is a view southwest towards Lee Mine, taken just south of Plate 31. The camera point was 0.8 km northeast of the intersection of sections 21 and 29, T1N R72W, at an elevation of 2501 m.

Original: c. 1902, J. P. Sturtevant No. 1729, BHS.
Match: 1984, T. T. Veblen and D. C. Lorenz No. B28.
Description: This is a closer view of the center background slopes shown in Plate 31. The lack of railroad tracks indicates that Plate 32 was taken several years before Plate 31. The Eldora line operated from 1903 to 1920, after which the tracks were removed. Since J. P. Sturtevant, the

Plate 32. Lee Mine. *Match*

photographer, died in 1910, this photograph must have been taken just before the laying of the rails. The match shows a dramatic increase in density of ponderosa pine forests and invasion into the grassland. A sample was taken a few meters above the railroad bed in the center of the photograph in the area of grassland invasion by trees. It indicates that 98 percent of the trees are ponderosa pine and 2 percent are Douglas fir and that the tree invasion began in the 1890s. Surrounding the invaded grassland site, ponderosa pine as old as 400 years were found. The upper left of the original photograph provides an example of the scale of nineteenth century burning and subsequent recovery.

Plate 33. Cardinal. *Original*

Location: This is an east-southeast view of Cardinal, located 2.5 km west of Nederland. The photo was taken 0.5 km southwest of the intersection of sections 11 and 15, T1S R73W, at an altitude of 2647 m.

Original: c. 1910, unknown photographer, Kemp (1960).

Match: 1985, T. T. Veblen and D. C. Lorenz No. B48.
Description: Cardinal was the site of an important tellurium mine in the 1890s. What was originally an open stand at the turn of the century is now a dense forest of mainly lodgepole pine. The former railroad bed is now covered by aspen.

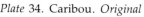

Location: This is a view southwest taken from 0.2 km west of the former town of Caribou, situated 8.5 km northwest of Nederland. The camera point was 0.5 km west-southwest of the intersection of sections 8 and 4, T1S R73W, at an elevation of 3105 m.
Original: c. 1905, J. P. Sturtevant 1438, BHS.
Match: 1985, T. T. Veblen and D. C. Lorenz No. B46.

Description: Caribou was originally a major silver mining town, established in 1869, and later was an important production center for tellurium. It was one of the highest-elevation settlements in the

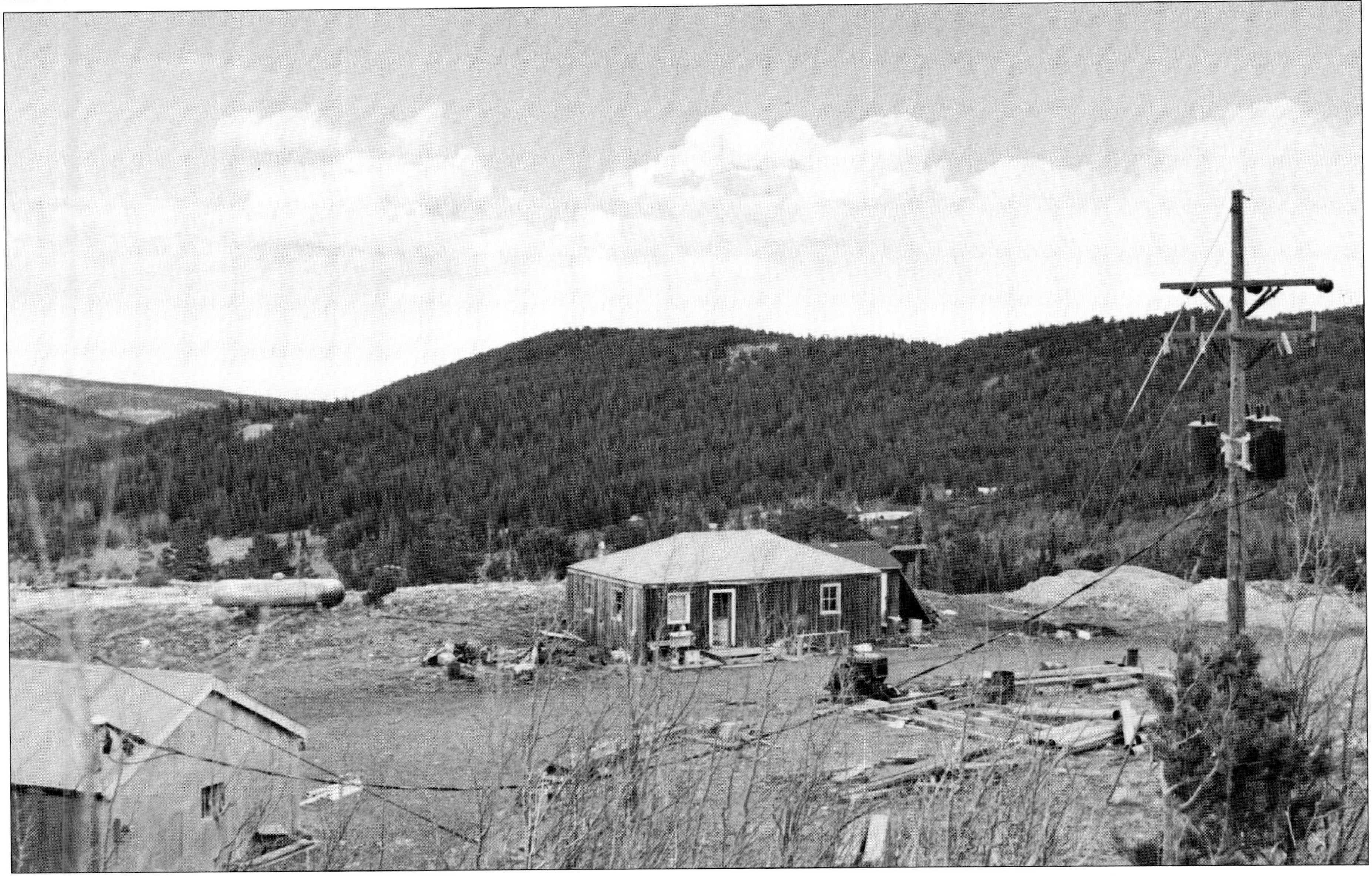

Plate 34. Caribou. *Match*

Front Range. The original photograph shows extensive logging around the settlement, especially in the right background. A sample of the present forest taken in this area indicates that limber pine accounts for 77 percent of the trees. Subalpine fir, Engelmann spruce, and quaking aspen are the other common species on this high elevation slope. The westerly slope and exposure to strong desiccating winds presumably account for the dominance by limber pine. Cut stumps and remnant trees over 200 years old indicate that limber pine also dominated the site prior to logging.

Plate 35. Eldora, east. *Original*

Location: This is a northwest view of Eldora Mountain taken from the town of Eldora, 5.0 km southwest of Nederland. The camera point was 0.5 km southeast of the intersection of sections 20 and 16, T1S R73W, at an elevation of 2660 m.
Original: 1898, J. P. Sturtevant No. 43209, CHS.
Match: 1985, T. T. Veblen and D. C. Lorenz No. B42.
Description: The discovery of gold in the mid-1890s

Plate 35. Eldora, east. *Match*

spawned the development of this boom town. The original photograph shows extensively burned forests, which appear to have been dominated by conifers, as judged from the shape of the snags and the few remnant trees. The match indicates the recovering stand is dominated mainly by quaking aspen (leafless) with scattered subalpine fir and Douglas fir.

Plate 36. Eldora, west. *Original*

Location: This is a view east of Eldora Mountain taken 0.5 km southeast of the intersection of sections 20 and 16, T1S R73W, at an elevation of 2660 m.

Original: 1897, J. P. Sturtevant No. 6581, CHS.
Match: 1985, T. T. Veblen and D. C. Lorenz No. B41.

Description: This site is adjacent to the site shown in Plate 35 and also shows an extensive area of burned forest. A sample of the present stand

Plate 36. Eldora, west. *Match*

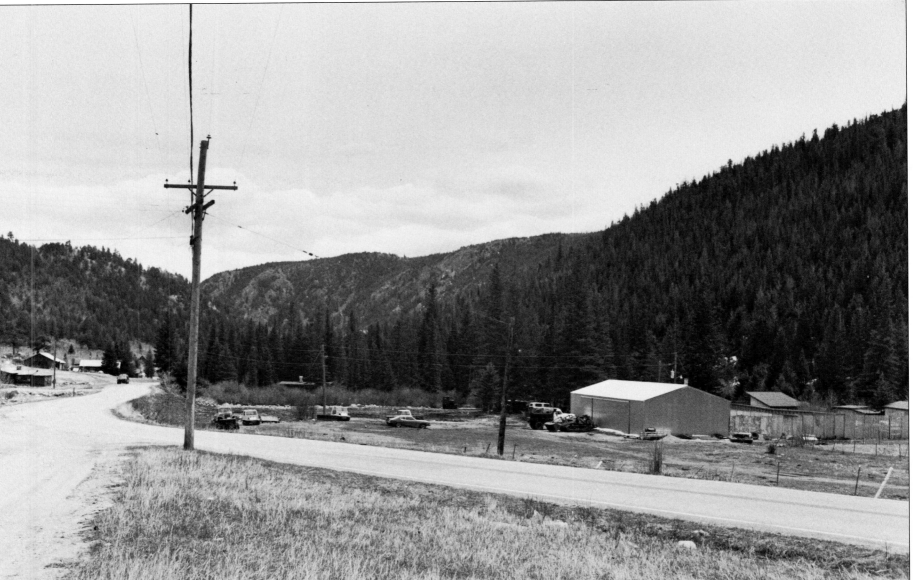

indicates recovery to subalpine fir (44 percent), Douglas fir (28 percent), Engelmann spruce (22 percent), limber pine (4 percent), and lodgepole pine (2 percent).

Plate 37. Clyde Mine. *Original*

Location: This is a view west of Clyde Mine, located 9.0 km north of Nederland on the Peak to Peak Highway (Highway 72). The camera point was 0.1 km south-southeast of the intersection of sections 1 and 7, T1S R72W, at an elevation of 2577 m.

Original: 1917, F. L. Hess No. 672, USGS.

Match: 1984, T. T. Veblen and D. C. Lorenz No. B5.

Description: This is a tungsten mine established in

1904. The original photograph shows selective logging of the surrounding forest for mine construction and fuel. The match shows recovery to a denser stand of lodgepole pine, Engelmann spruce, and subalpine fir.

Plate 38. Gale Mine. *Original*

Location: This view west-southwest of Gale Mine, located west of Clyde Mine, was taken 0.1 km south-southwest of the intersection of sections 1 and 7, T1S R72W, at an elevation of 2577 m.

Original: 1917, F. L. Hess No. 671, USGS.
Match: 1984, T. T. Veblen and D. C. Lorenz No. B6.

Plate 38. Gale Mine. *Match*

Description: The original photograph shows logging, particularly of the left slope, of the area surrounding the mine. The match shows recovery to lodgepole pine and aspen. Gale Mine is still active.

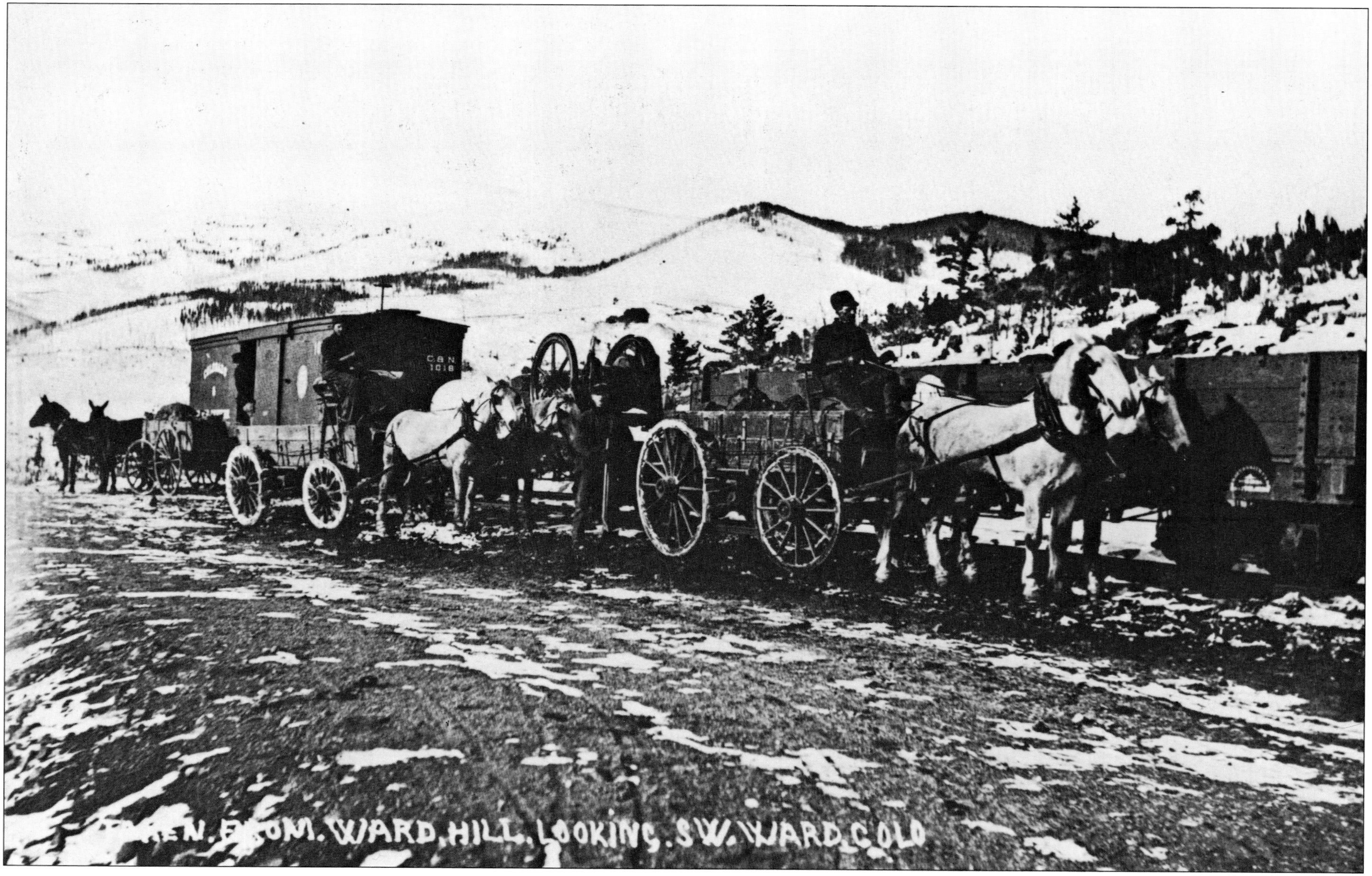

Plate 39. New Market. *Original*

Location: This view southwest from New Market is located at the northeast corner of Ward along the Peak to Peak Highway. The camera point was 0.2 km west of the intersection of sections 6 and 12, T1N R73W, at an elevation of 2837 m.
Original: 1905 Kindig Collection, Crossen (1978).

Plate 39. New Market. *Match*

Match: 1982, N. R. Johnson No. B67.
Description: The original photograph shows an extensive area of burned forest at the turn of the century. The match shows recovery to dense stands of mainly lodgepole pine.

Plate 40. Upper Lefthand Canyon. *Original*

Location: This north-northwest view of Lefthand Canyon was taken from the town of Puzzler. The camera point was 0.8 km west-southwest of the intersection of sections 8 and 18, T1N R72W, at an altitude of 2715 m.

Original: 1900, unknown photographer, BHS.

Match: 1982, N. R. Johnson No. B68.

Description: The original photograph shows the extent of burned forests in upper Lefthand Canyon

Plate 40. Upper Lefthand Canyon. *Match*

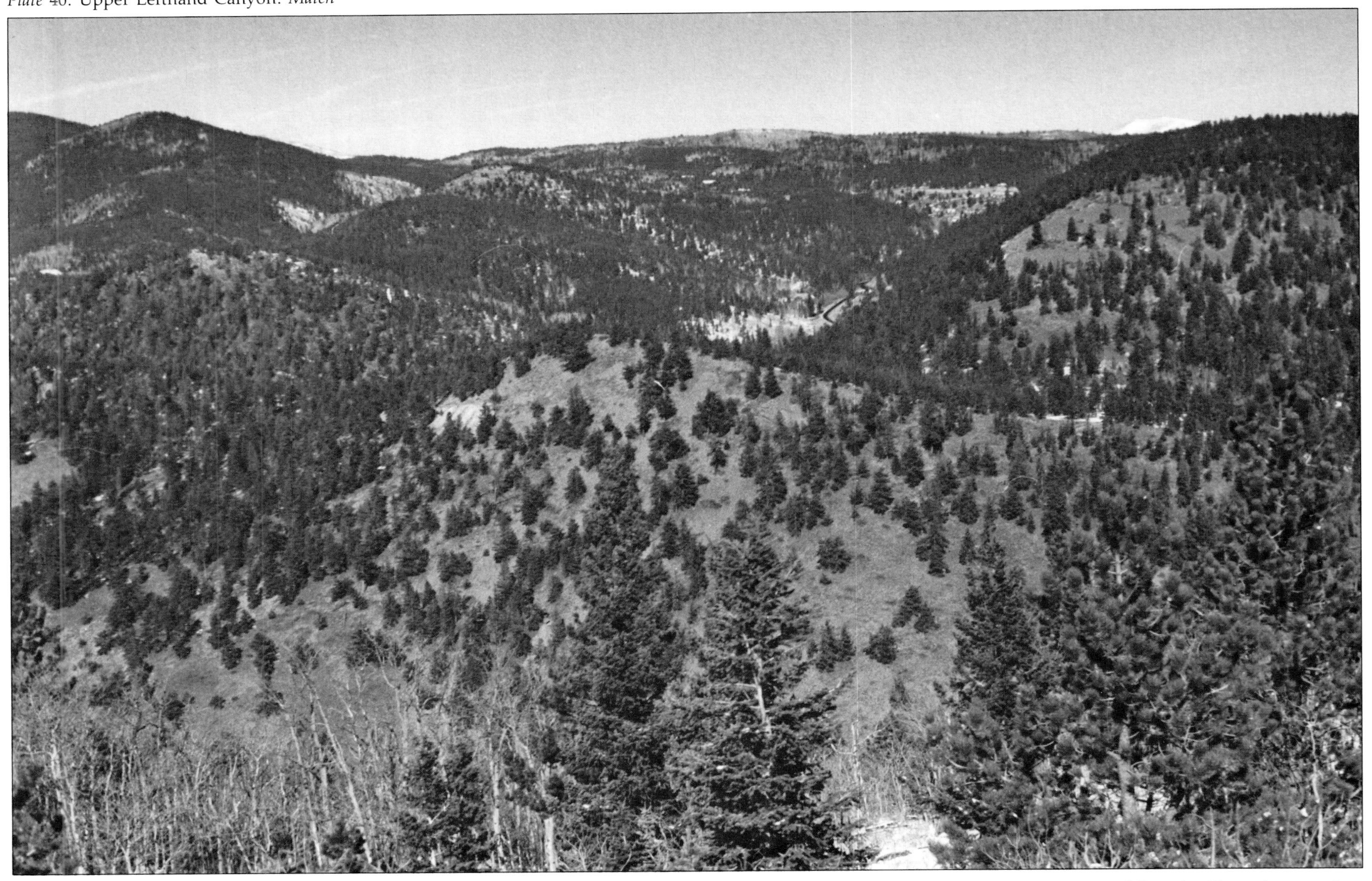

at the turn of the century. This was an area of numerous mines such as Utica, Superior, Morning Star, and Mountain Chief mines. The match shows recovery to stands on the lower slopes dominated by ponderosa pine and Douglas Fir and to stands on the upper slopes dominated by lodgepole pine.

Plate 41. Loder Smelter, view northwest. *Original*

Location: This is a northwest view of the Loder Smelter, located 2.25 km south of Ward. The camera point was 0.25 km southeast of the intersection of sections 7 and 13, T1N R73W, at an elevation of 2599 m.
Original: c. 1905, J. P. Sturtevant No. 1828, BHS.

Match: 1984, T. T. Veblen and D. C. Lorenz No. B23.
Description: Burned spars are common on the left

Plate 41. Loder Smelter, view northwest. *Match*

slope and in the center-background of the original photograph. The foreground had been largely cleared, as indicated by numerous cut stumps. The match shows a substantial increase in Douglas fir density on the northerly slope to the left and recovery of the cleared site to an aspen-dominated stand.

Plate 42. Loder Smelter, view south. *Original*

Location: This photo was taken 0.7 km west-southwest of the intersection of sections 8 and 18, T1N R73W, at an elevation of 2599 m. See Plate 41 for general location.

Original: c. 1905, J. P. Sturtevant No. 1847, BHS. *Match:* 1984, T. T. Veblen and D. C. Lorenz No. B22.

Description: The Loder Smelter was operated by the Ward Pyritic Smelter Company and at the turn of the century served Ward, Puzzler, and the Big Five

Plate 42. Loder Smelter, view south. *Match*

mines. The original photo shows that most of this north-facing slope had been burned and cleared. A sample from the center of the photograph indicates that today the forest is over 50 percent Douglas fir mixed with smaller numbers of lodgepole pine, Engelmann spruce, limber pine, and subalpine fir. The gray tones in the midground of the modern photograph are the crowns of Douglas fir defoliated by spruce budworm.

Plate 43. Puzzler. *Original*

Location: This is a view northwest from Puzzler, 3.0 km south of Ward. The photo was taken 0.6 km south-southwest of the intersection of sections 13 and 7, T1N R73W, at an elevation of 2669 m.
Original: 1910, Kindig Collection, Crossen (1978).

Match: 1982, N. R. Johnson No. B58.
Description: Puzzler was a gold-mining settlement established in the 1890s. In the original photo-

graph, lodgepole pine and limber pine forests at the town site are cut, but seedlings and saplings are common. In the left midground there has been recovery to a dense stand of lodgepole pine. Tree density on the right background slopes has also substantially increased.

Plate 44. Big Five Mine. *Original*

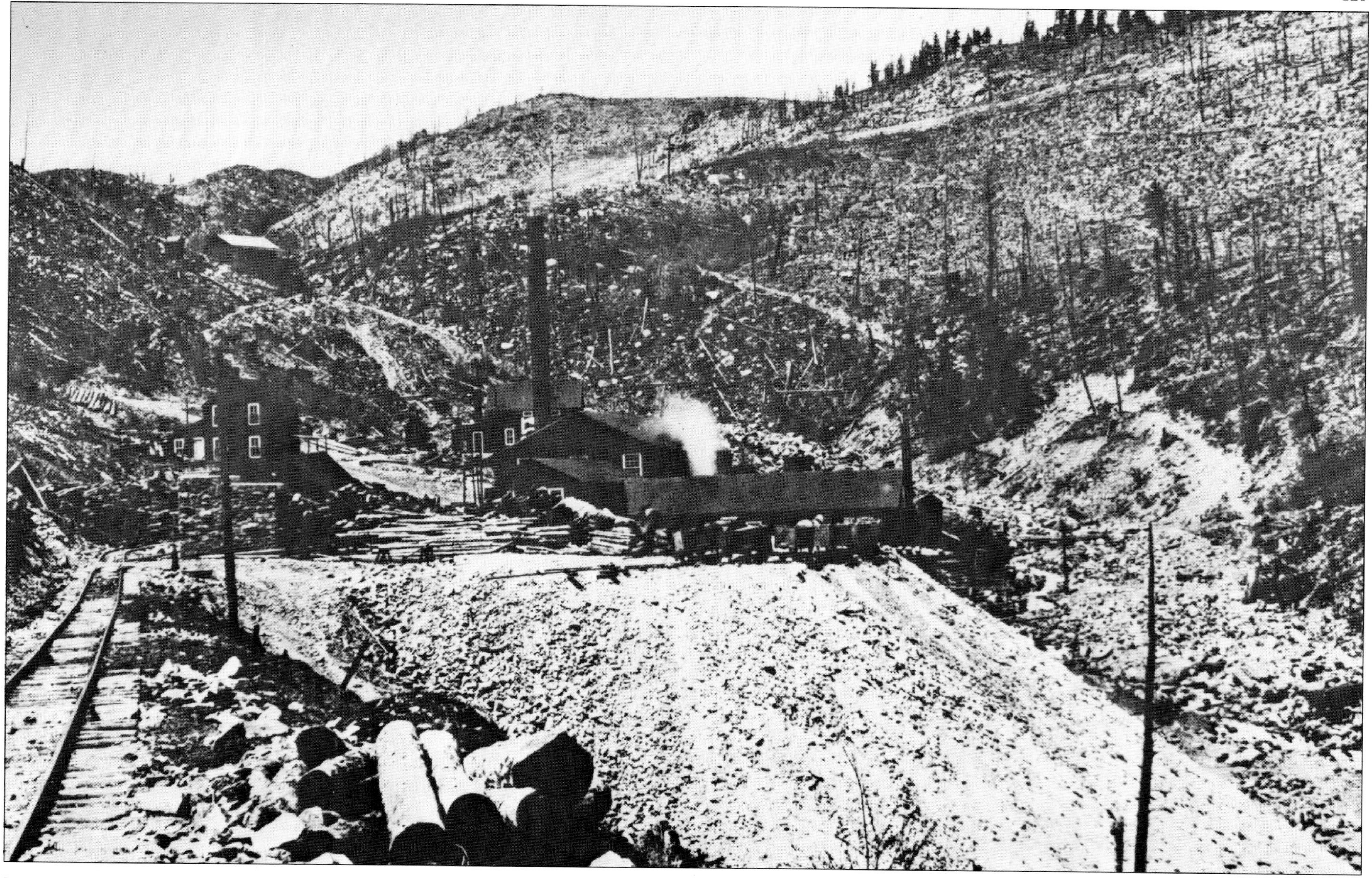

Location: This is a northwest view of Big Five Mine, situated 1.5 km south of Ward along the Ward line of the Switzerland Trail. The camera point was 1.0 km southeast from the intersection of sections 1 and 11, T1N R73W, at an elevation of 2678 m.
Original: 1905, Kindig Collection, Crossen (1978).

Match: 1982, N. R. Johnson No. B21.
Description: The original photograph shows the devastation caused by logging and burning of the

Plate 44. Big Five Mine. *Match*

slopes surrounding this important mine. The slope in the upper right was sampled and consisted of 67 percent limber pine, 22 percent ponderosa pine, 6 percent Douglas fir, and 6 percent aspen. The composition of remnant trees and cut stumps indicates that there has been a postdisturbance shift in composition from Douglas fir toward limber pine.

Plate 45. Twin Sister Peaks. *Original*

Location: This is a view north of Estes Park from the summit of Twin Sister Peaks. The camera point was 0.3 km northeast of the intersection of sections 26 and 24, T4N R73W, at an elevation of 3486 m. *Original:* c. 1916, W. T. Lee No. 1067, USGS. *Match:* 1986, T. T. Veblen and D. C. Lorenz No. R17.

Plate 45. Twin Sister Peaks. *Match*

Description: The original photograph shows a large opening in the center foreground that appears to have been burned. The match indicates a substantial recovery to a dense stand of subalpine fir, Engelmann spruce, and lodgepole pine. The distant background slopes generally support denser tree covers than they did at the time of the original photograph.

Plate 46. Marys Lake. *Original*

Location: This is a view northwest, with Marys Lake in the foreground and the Continental Divide in the background. The photo was taken 0.9 km northwest of the intersection of sections 11 and 1, T4N R73W, at an altitude of 2654 m. *Original:* c. 1915, L. C. McClure, DPL.

Plate 46. Marys Lake. *Match*

Match: 1986, T. T. Veblen and D. C. Lorenz No. L12.

Description: Marys Lake receives water from the Colorado River via a transmontane aqueduct. The match shows an increase in ponderosa pine density along the far shore.

Plate 47. Lake Estes and Mt. Olympus. *Original*

Location: This is a view east of Lake Estes and Mt. Olympus taken from present-day Estes Park. The camera point was 0.4 km southwest of the intersection of sections 25 and 19, T5N R73W, at an elevation of 1739 m.
Original: c. 1915, L. C. McClure No. 1819, DPL.

Match: 1986, T. T. Veblen and D. C. Lorenz No. L16.

Plate 47. Lake Estes and Mt. Olympus. *Match*

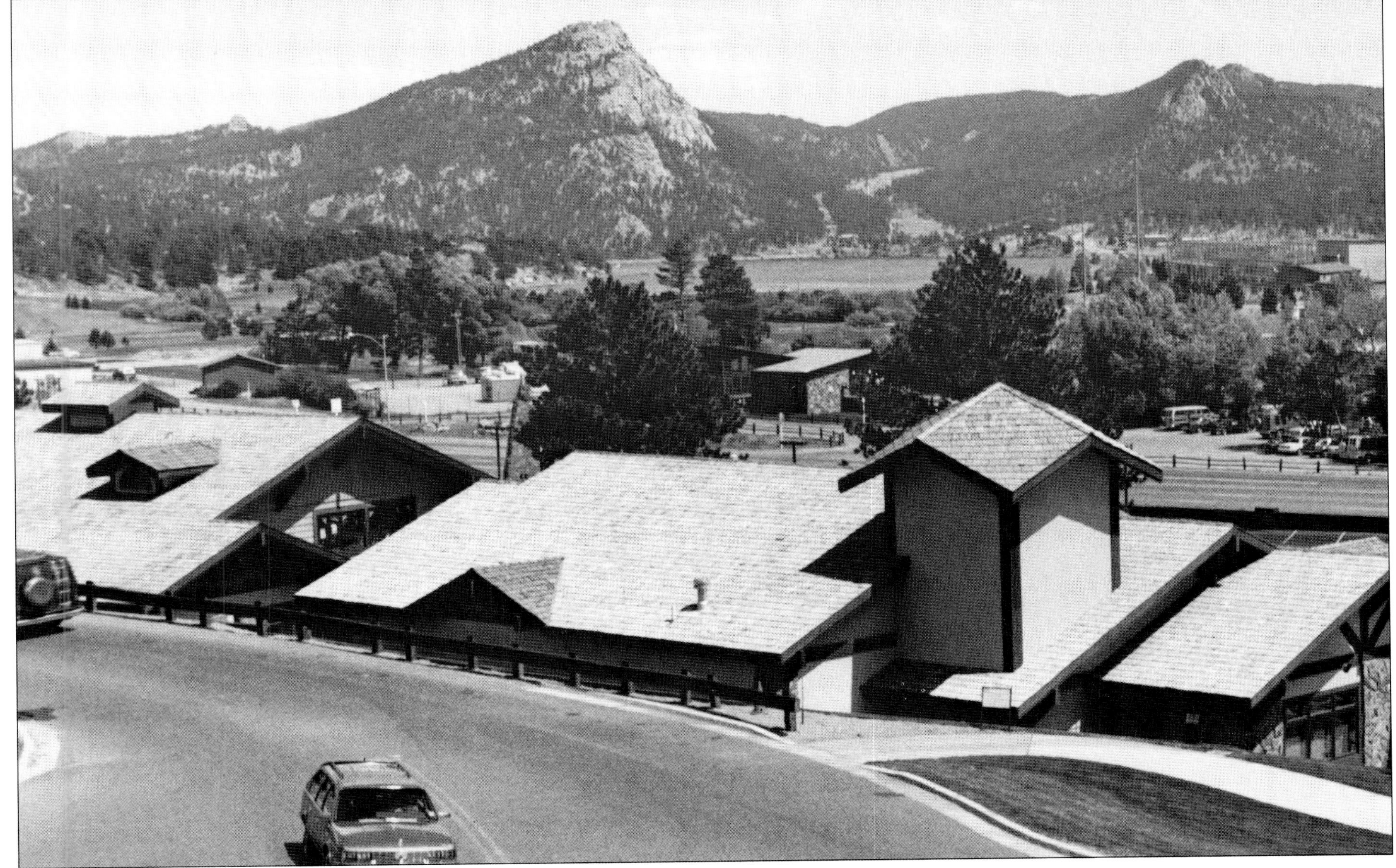

Description: Lake Estes was formed by the damming of the Big Thompson River. The main change shown by the match is the urban development of the eastern edge of the resort town of Estes Park. The forests of Douglas fir and ponderosa pine in the background have also increased in density.

Plate 48. Rams Horn Mountain from Prospect Mountain. *Original*

Location: This is a view to the southwest of Marys Lake and Rams Horn Mountain. The photo was taken 0.4 km west of the intersection of sections 36 and 30, T5N R73W, at an elevation of 2623 m. *Original:* 1921, W. T. Lee No. 2036, USGS.

Match: 1986, T. T. Veblen and D. C. Lorenz No. L1.

Plate 48. Rams Horn Mountain from Prospect Mountain. *Match*

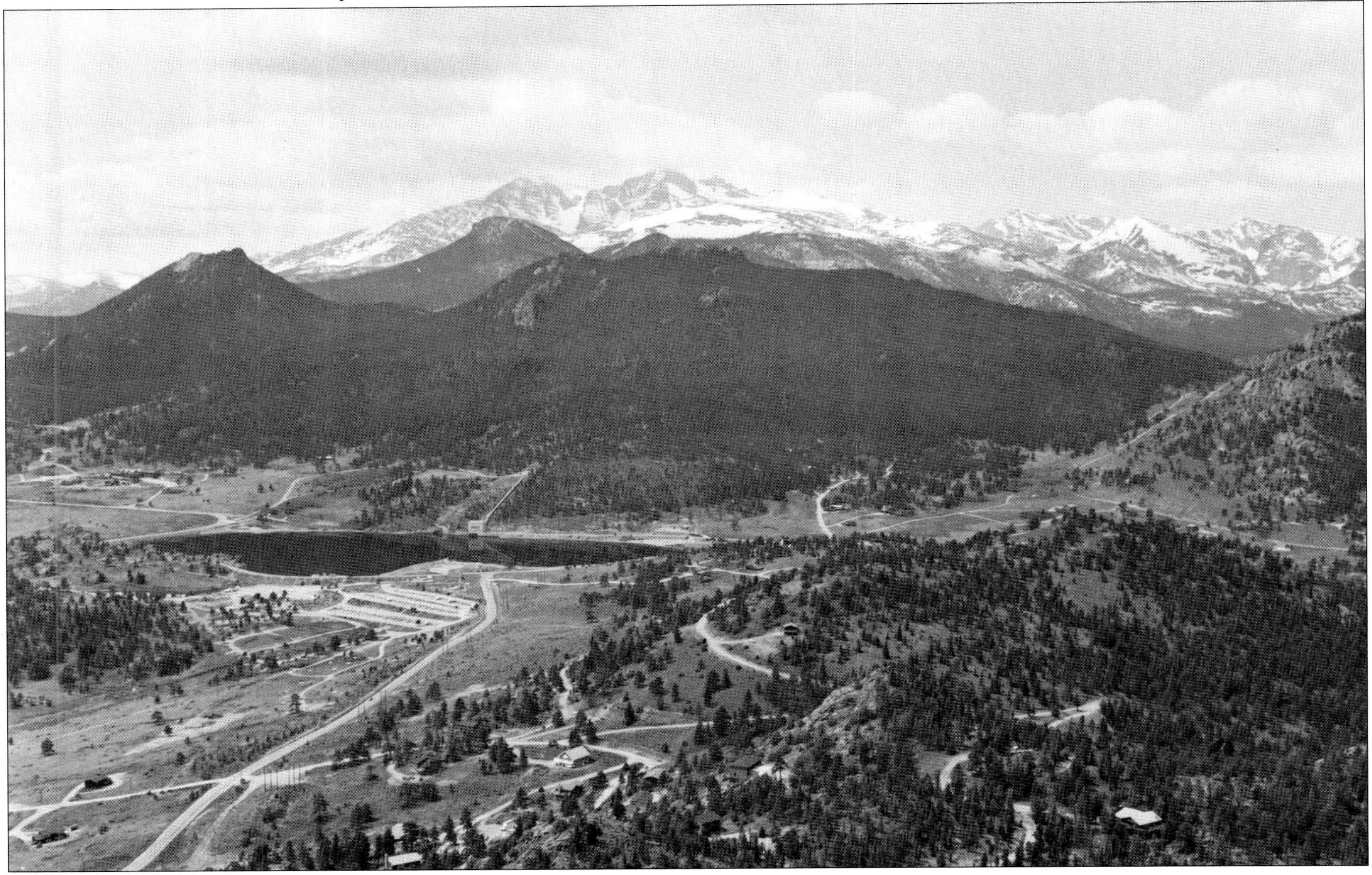

Description: The right side of this scene is the same as center and left foreground of Plate 49. Again, the match shows an increase in the density of ponderosa pine stands. Marys Lake has expanded from a small pond to a larger reservoir and is bordered towards the foreground by a large campground.

Plate 49. Beaver Mountain from Prospect Mountain. *Original*

Location: This is a west view of Beaver Mountain taken from Prospect Mountain. The camera point was 0.4 km west of the intersection of sections 36 and 30, T5N R73W, at an elevation of 2623 m. *Original:* 1921, W. T. Lee No. 2037, USGS.

Match: 1986, T. T. Veblen and D. C. Lorenz No. L2.

Plate 49. Beaver Mountain from Prospect Mountain. *Match*

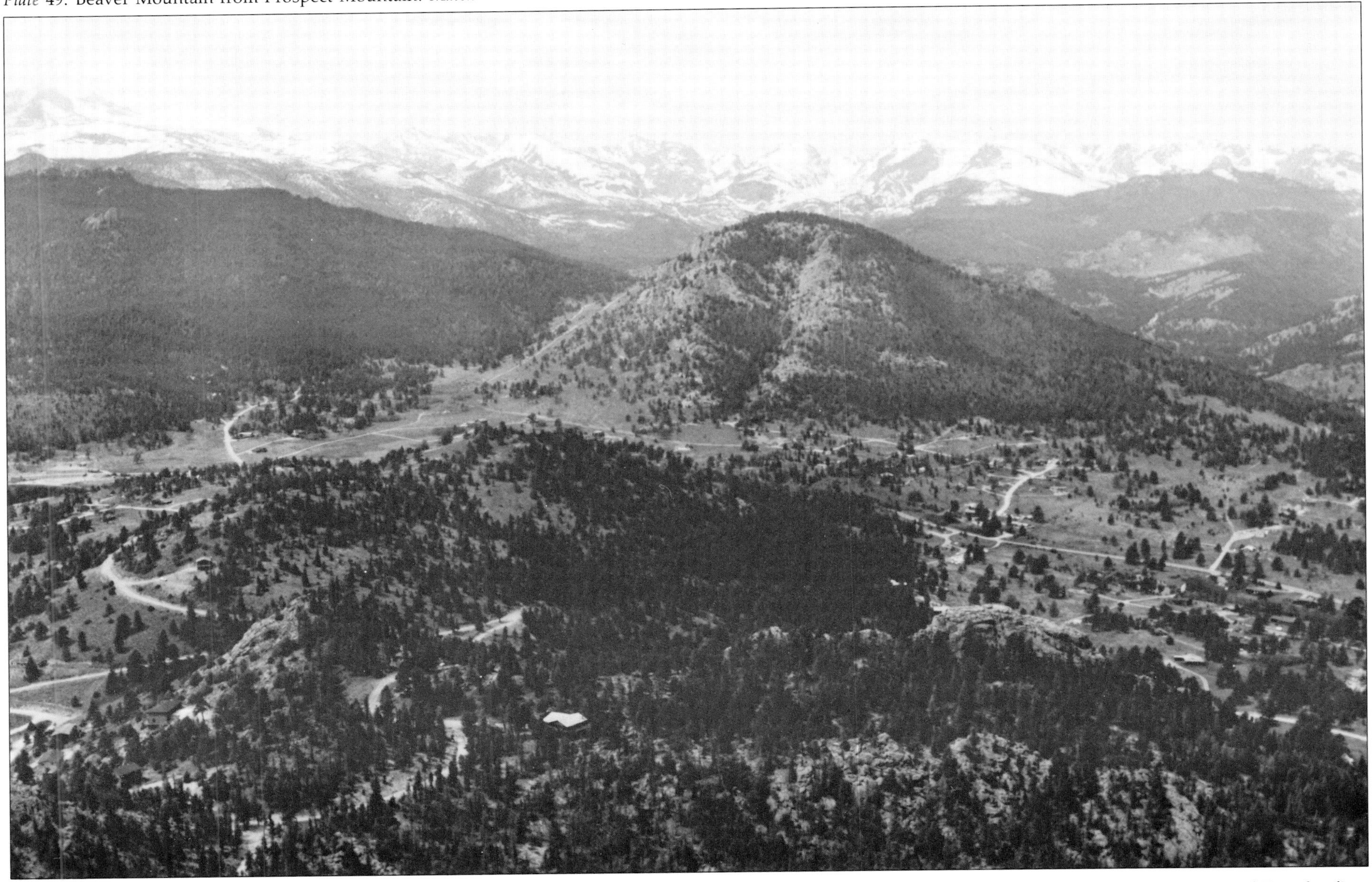

Description: The open ponderosa pine woodlands of the center and foreground in the old photograph appear to have been logged. The match shows a marked increase in tree density, particularly of Douglas fir. The gray tones in the right center of the present photograph are crowns of Douglas fir defoliated by spruce budworm.

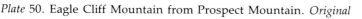

Plate 50. Eagle Cliff Mountain from Prospect Mountain. *Original*

Location: This west-northwest view from Prospect Mountain was taken 0.4 km west of the intersection of sections 36 and 30, T5N R73W, at an elevation of 2623 m.

Original: 1921, W. T. Lee No. 2038, USGS.
Match: 1986, T. T. Veblen and D. C. Lorenz No. L3.

Plate 50. Eagle Cliff Mountain from Prospect Mountain. *Match*

Description: This is a view of the main entrance to Rocky Mountain National Park. The main changes are the increase in density of ponderosa pine stands and suburban development on the western side of Estes Park. Spruce budworm defoliation of Douglas fir is noticeable in the left center.

Plate 51. Deer Mountain from Prospect Mountain. *Original*

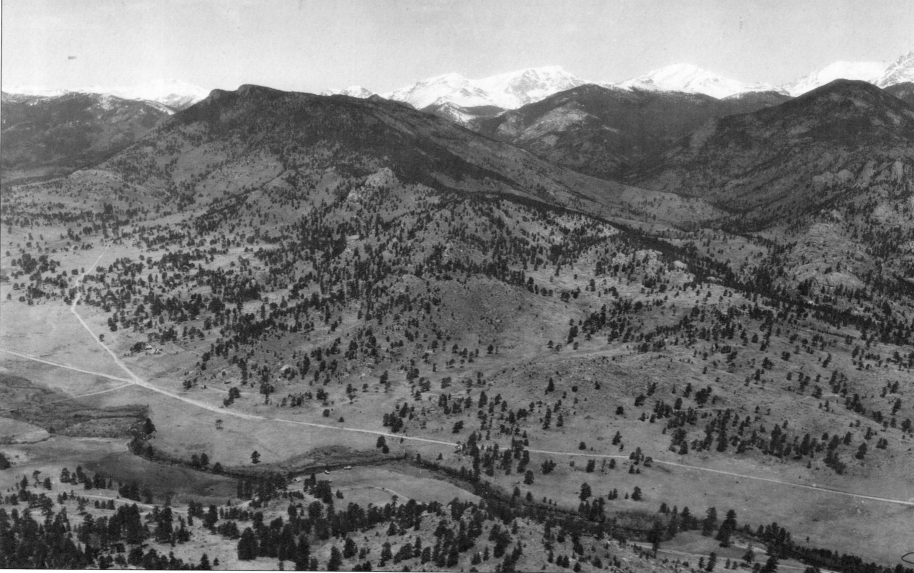

Location: This northwest view from Prospect Mountain was taken 0.4 km west of the intersection of sections 36 and 30, T5N R73W, at an elevation of 2623 m.

Original: 1921, W. T. Lee No. 2039, USGS.

Plate 51. Deer Mountain from Prospect Mountain. *Match*

Match: 1986, T. T. Veblen and D. C. Lorenz No. L4.

Description: This view overlaps with the right side of Plate 50. Again, the match shows an increase in the density of ponderosa pine woodlands and considerable suburban development.

Plate 52. Twin Owls from Devils Gulch Road. *Original*

Location: This view north of the Twin Owls and Lumpy Ridge was taken 0.9 km northeast of the intersection of sections 26 and 24, T5N R73W, at an elevation of 2318 m.

Original: c. 1900, unknown photographer, DPL.

Plate 52. Twin Owls from Devils Gulch Road. *Match*

Match: 1986, T. T. Veblen and D. C. Lorenz No. L18.

Description: In the original photograph the rocky slopes in the background are sparsely covered with ponderosa pine and Douglas fir. The match shows a dramatic increase in stand density. Gray tones in the background are crowns of Douglas fir recently defoliated by spruce budworm.

Plate 53. Longs Peak Inn and Mills Moraine. *Original*

Location: This is a southwest view of Longs Peak and Mills Moraine, taken 0.6 km southeast of the intersection of sections 3 and 35, T4N R73W, at an elevation of 2660 m.

Original: 1900, Betts, DPL.
Match: 1986, T. T. Veblen and D. C. Lorenz No. L9.

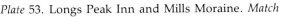

Plate 53. Longs Peak Inn and Mills Moraine. *Match*

Description: The original photograph shows extensive areas of burned and partially logged subalpine forest. The match indicates recovery to dense stands of mainly lodgepole pine with smaller numbers of Engelmann spruce, subalpine fir, and limber pine. The principal cultural change is the replacement of the small ranch with a resort hotel.

Plate 54. Longs Peak, Mills Moraine, and Pine Ridge. *Original*

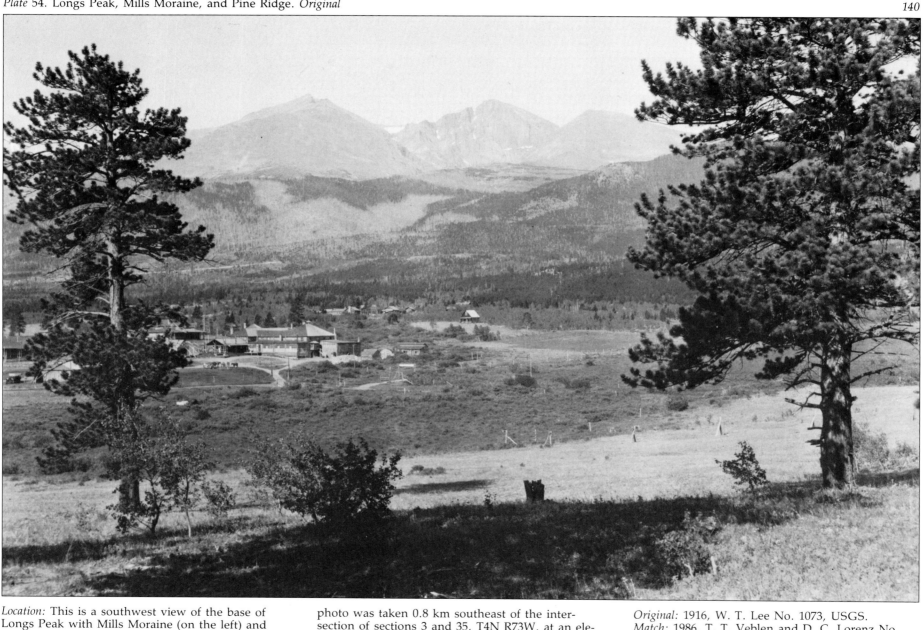

Location: This is a southwest view of the base of Longs Peak with Mills Moraine (on the left) and Pine Ridge (on the right) in the midground. The photo was taken 0.8 km southeast of the intersection of sections 3 and 35, T4N R73W, at an elevation of 2672 m.

Original: 1916, W. T. Lee No. 1073, USGS.
Match: 1986, T. T. Veblen and D. C. Lorenz No. L10.

Plate 54. Longs Peak, Mills Moraine, and Pine Ridge. *Match*

Description: This is a more distant view of much of the area seen in Plate 53, taken 16 years later. The original photograph shows that a sizable area of forest burned between 1900 and 1916. Recovery has been to lodgepole pine-dominated forest.

Plate 55. Longs Peak from Allenspark. *Original*

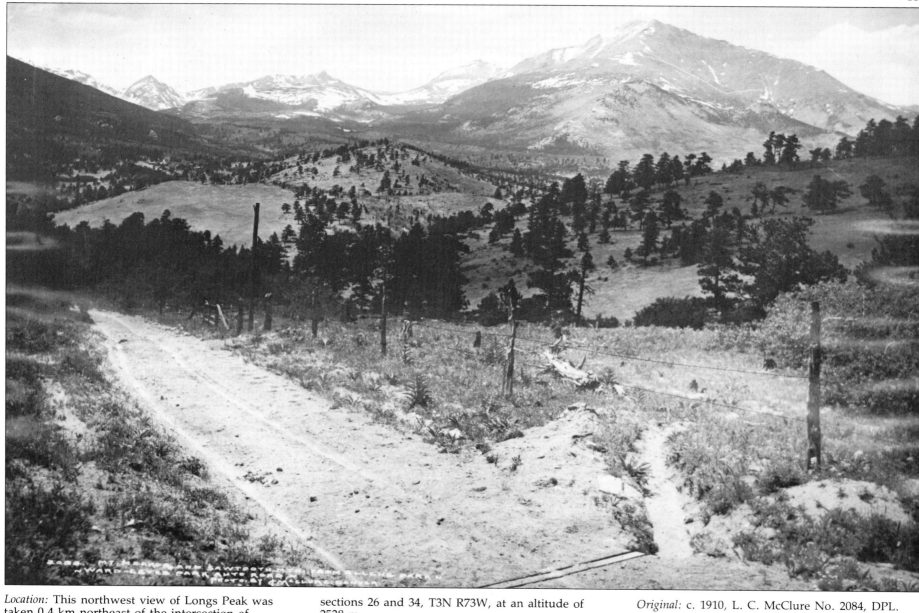

Location: This northwest view of Longs Peak was taken 0.4 km northeast of the intersection of sections 26 and 34, T3N R73W, at an altitude of 2538 m.

Original: c. 1910, L. C. McClure No. 2084, DPL.

Plate 55. Longs Peak from Allenspark. *Match*

Match: 1986, T. T. Veblen and D. C. Lorenz No. L15.
Description: On both sides of the ridge below Longs Peak (upper right) extensive burns are visible in the original photograph. The match shows dramatic recovery of these burns to lodgepole pine–dominated stands. In the foreground, ponderosa pine has increased in density, yet other meadows (perhaps still grazed) remain free of trees.

Plate 56. Copeland Mountain and Wild Basin. *Original*

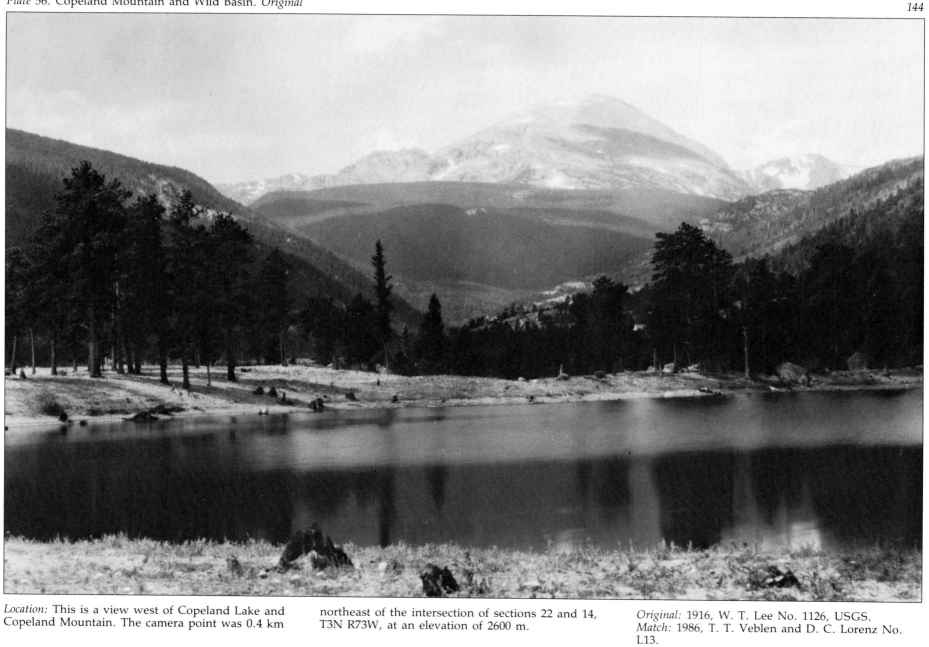

Location: This is a view west of Copeland Lake and Copeland Mountain. The camera point was 0.4 km northeast of the intersection of sections 22 and 14, T3N R73W, at an elevation of 2600 m.

Original: 1916, W. T. Lee No. 1126, USGS.
Match: 1986, T. T. Veblen and D. C. Lorenz No. L13.

Plate 56. Copeland Mountain and Wild Basin. *Match*

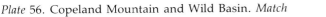

Description: Burns are evident on the middle slopes in the center background of the original photograph. The 1978 Ouzel burn can be seen in the same general location on the modern photograph. The match also shows an increase in shrubs along the margin of the lake.

Plate 57. Hallett Peak and Bear Lake. *Original*

Location: This photograph was taken from the east shore of Bear Lake, 4.5 km south-southwest of the intersection of sections 36 and 31, T4N R74W, at an elevation of 2880 m.

Original: 1916, W. T. Lee No. 1165, USGS.
Match: 1986, T. T. Veblen and D. C. Lorenz No. R9.

Description: The original photo shows that the southern and western side of Bear Lake had been burned during the nineteenth century. By the time

Plate 57. Hallett Peak and Bear Lake. *Match*

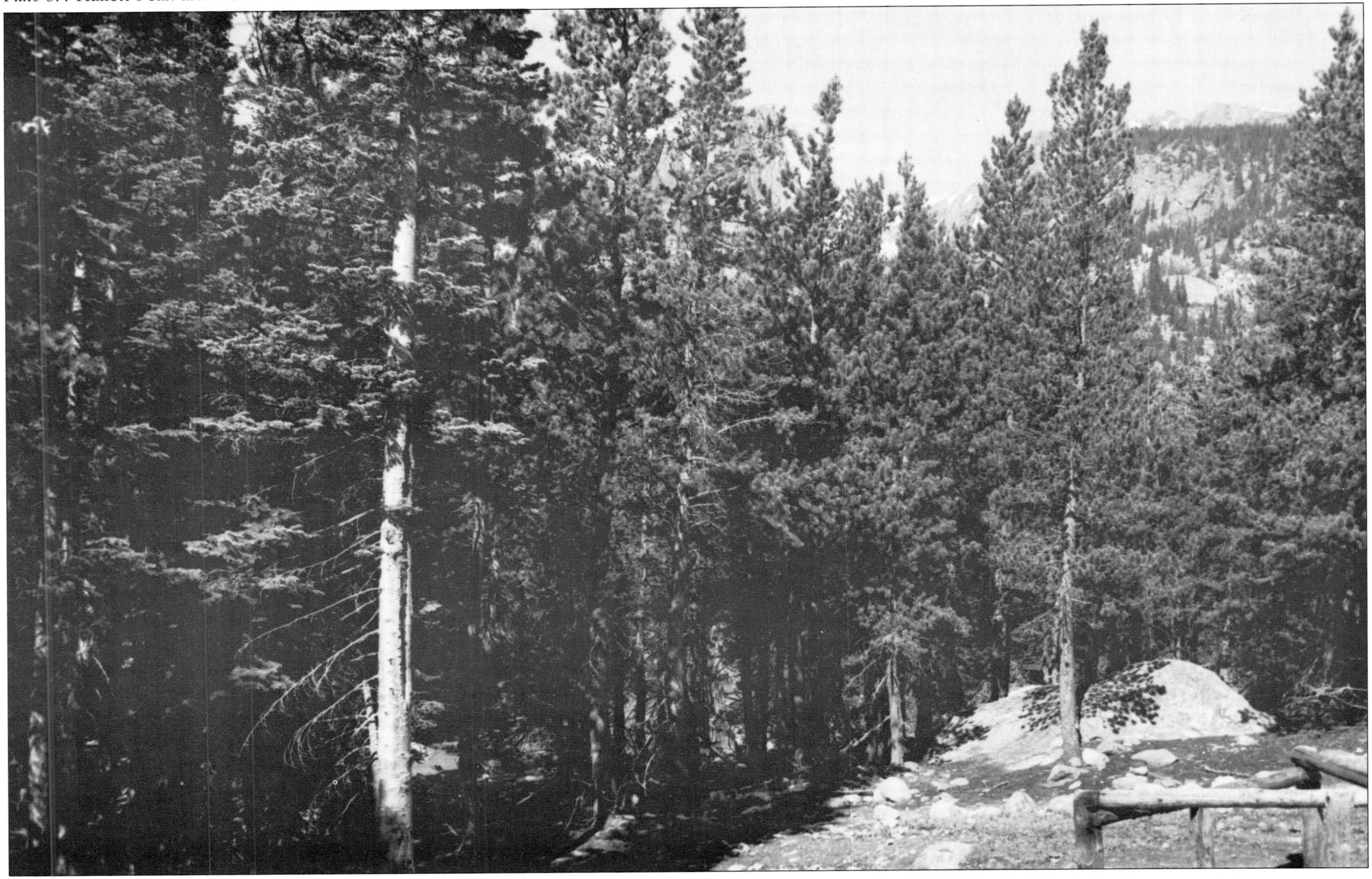

of the 1916 photograph, small lodgepole pine and Engelmann spruce are abundant, as seen in the foreground. In the modern photograph lodgepole pine in the foreground obscures most of the scene visible in the original photograph. The obscured background slopes are now postfire stands of mainly lodgepole pine with smaller numbers of Engelmann spruce and subalpine fir.

Plate 58. Bear Lake. *Original*

Location: This western view of the Lake and Hallett Peak was taken 4.5 km south-southwest of the intersection of sections 36 and 31, T4N R74W, at an elevation of 2880 m.

Original: c. 1916, W. T. Lee No. 1179, USGS.
Match: 1986, T. T. Veblen and D. C. Lorenz No. R5.

Plate 58. Bear Lake. *Match*

Description: This view overlaps considerably with the view in Plate 59. The match shows postfire recovery to stands of lodgepole pine, Engelmann spruce, and subalpine fir on the far side of the lake. Limber pine also occurs here on the rockiest sites.

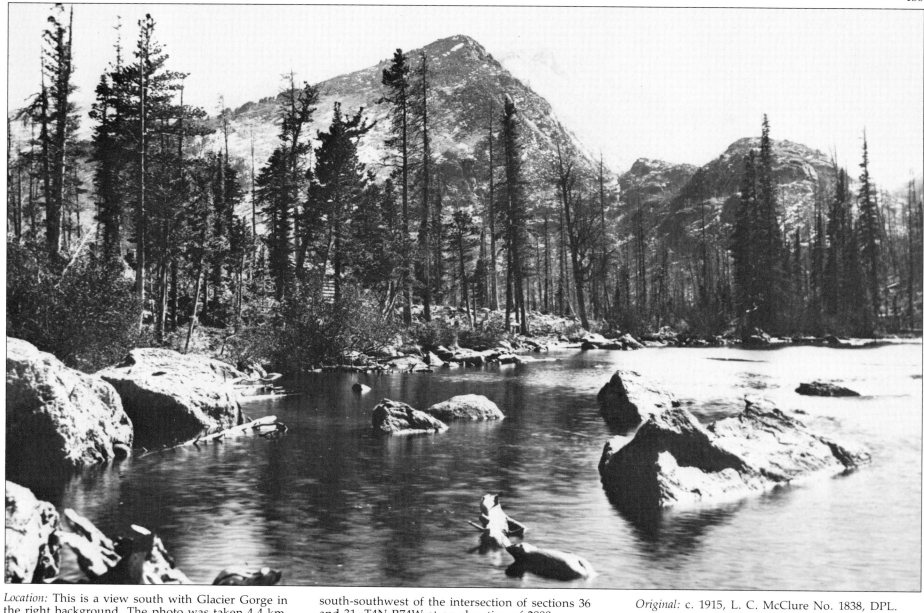

Plate 59. Bear Lake and Half Mountain. *Original*

Location: This is a view south with Glacier Gorge in the right background. The photo was taken 4.4 km south-southwest of the intersection of sections 36 and 31, T4N R74W at an elevation of 2880 m.

Original: c. 1915, L. C. McClure No. 1838, DPL.

Plate 59. Bear Lake and Half Mountain. *Match*

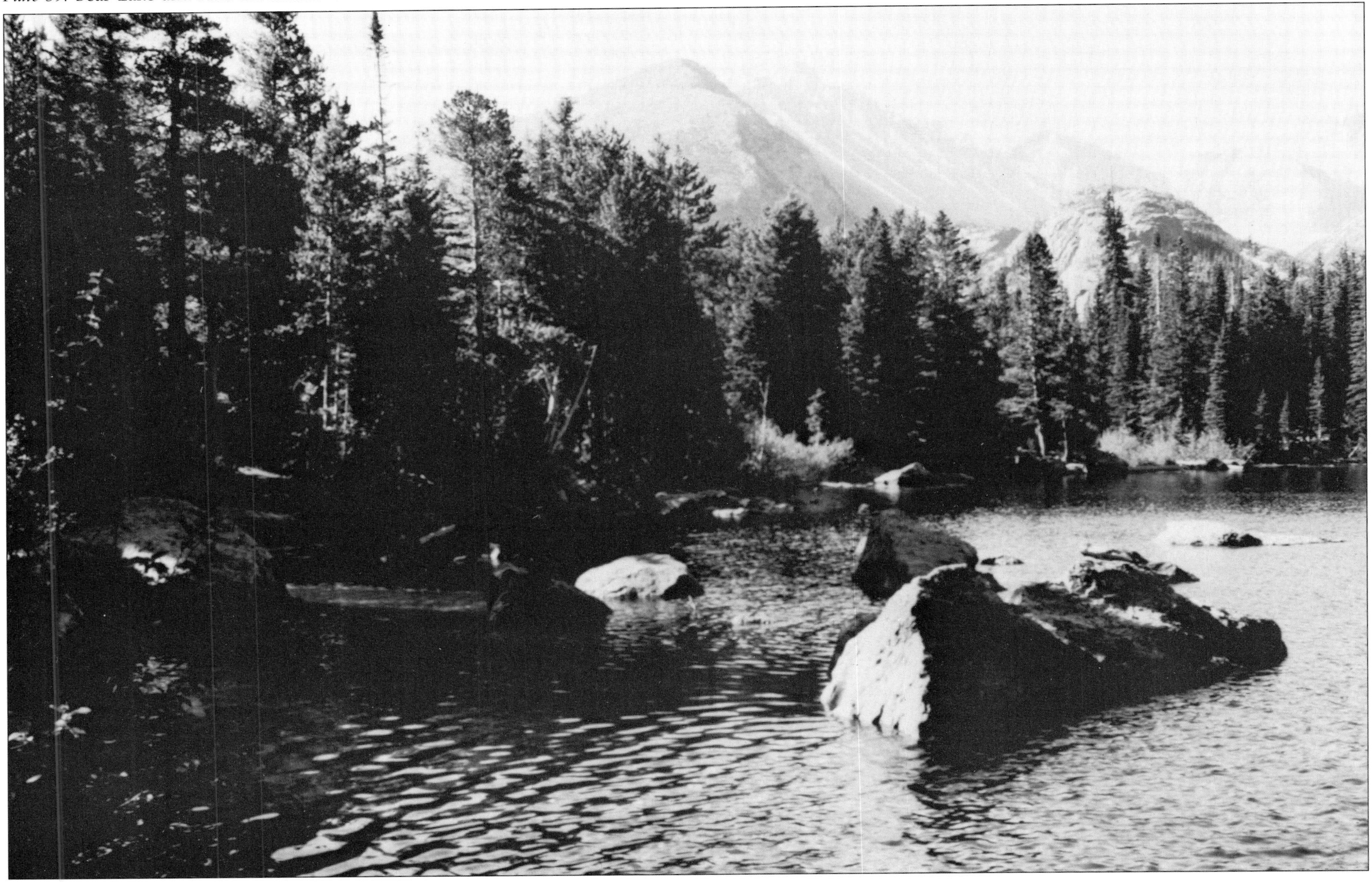

Match: 1986, T. T. Veblen and D. C. Lorenz R13.

Description: In the original photograph, the forests at the southern end of Bear Lake clearly have been burned. Some logging, probably for the construction of the cabin in the left center, was indicated by cut stumps. The match shows recovery to a forest dominated by lodgepole pine.

Plate 60. Lilypad Lake (now called Nymph Lake). *Original*

Location: This is a view to the northeast taken 4.9 km south-southwest of the intersection of sections 36 and 31, T4N R74W, at an elevation of 2965 m. *Original:* c. 1916, W. T. Lee No. 1133, USGS. *Match:* 1986, T. T. Veblen and D. C. Lorenz No. R16.

Plate 60. Lilypad Lake (now called Nymph Lake). *Match*

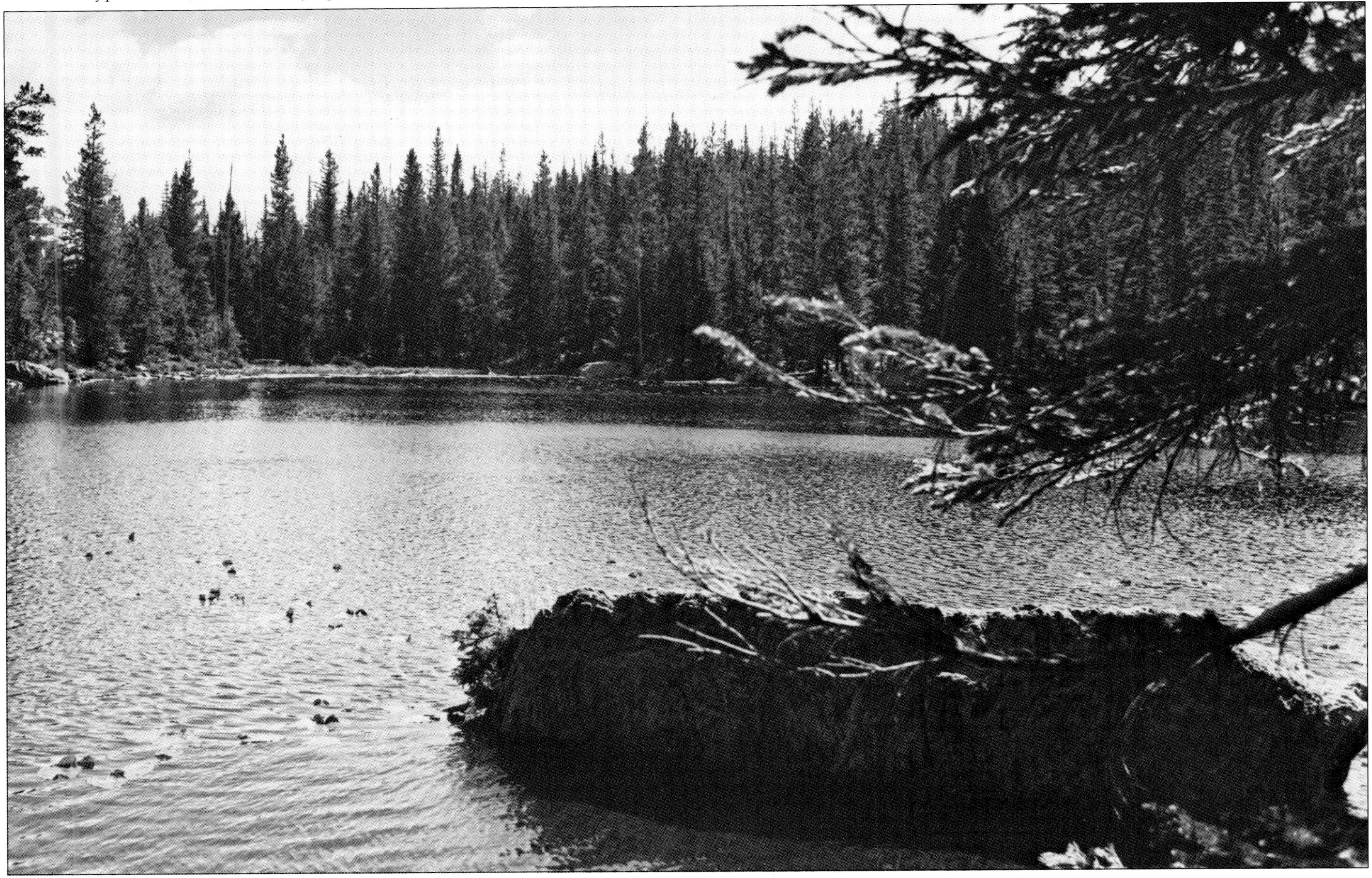

Description: The original shows that nearly all the forest on the western side of the lake had been burned. The match shows recovery to a dense stand of Engelmann spruce, subalpine fir, and lodgepole pine.

Plate 61. Bierstadt Lake, view north. *Original*

Location: This is a north-northeast view of Bierstadt Lake with Mount Wuh and Joe Mills Mountain in the background. The photo was taken 0.7 km south-southwest of the intersection of sections 6 and 7, T4N R74W, at an elevation of 2855 m.

Original: c. 1915, L. C. McClure No. 1835, DPL.

Plate 61. Bierstadt Lake, view north. *Match*

Match: 1986, T. T. Veblen and D. C. Lorenz No. R8.
Description: The original shows no evidence of recent fire on the northeastern side of Bierstadt Lake. The match indicates no major changes in this forest of Engelmann spruce, subalpine fir, and lodgepole pine.

Plate 62. Bierstadt Lake, view northwest. *Original*

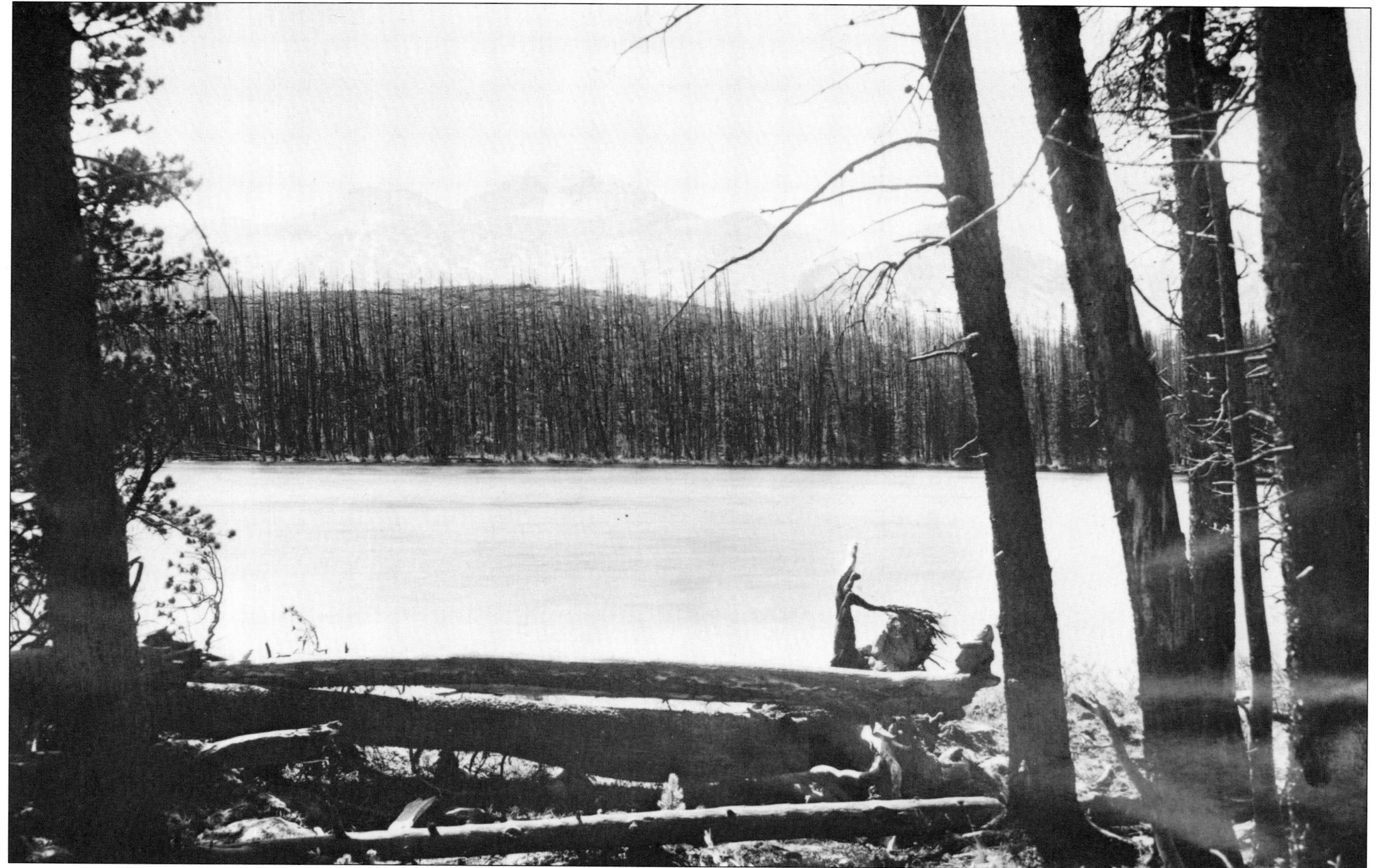

Location: This view to the south shows Longs Peak in the background. The photo was taken 0.7 km south-southwest of the intersection of sections 6 and 7, T4N R74W, at an elevation of 2855 m.

Original: c. 1915, L. C. McClure No. 1834, DPL.

Plate 62. Bierstadt Lake, view northwest. *Match*

Match: 1986, T. T. Veblen and D. C. Lorenz No. R7.

Description: The original, taken at the same time as the original for Plate 61 shows extensive burning of the subalpine forest on the southwestern side of Bierstadt Lake. Charcoal and burned spars are still common today. The match shows recovery to a dense stand of lodgepole pine.

Plate 63. The Loch. *Original*

Location: This west-southwest view of the lake and the Sharkstooth was taken 6.9 km south-southwest of the intersection of sections 31 and 36, T4N R74W, at an elevation of 3172 m.

Original: 1916, W. T. Lee No. 1151, USGS.
Match: 1986, T. T. Veblen and D. C. Lorenz No. R20.

Plate 63. The Loch. *Match*

Description: The original photograph shows evidence of small, patchy fires, but most of the basin was relatively undisturbed. Given the high elevation, these fires are more likely to have been lightning-ignited rather than human-set. The match indicates that the burned patches have recovered to stands of Engelmann spruce and subalpine fir.

Plate 64. Horseshoe Lake (now called Sheep Lake). *Original*

Location: This northwest view of Mt. Chapin, Mt. Chiquita, and Ypsilon Mountain was taken 1.0 km northwest of the intersection of sections 18 and 20, T5N R73W, at an elevation of 2593 m.
Original: c. 1915, L. C. McClure No. 2630, DPL.
Match: 1986, T. T. Veblen and D. C. Lorenz No. R21.
Description: These two photographs show a

Plate 64. Horseshoe Lake (now called Sheep Lake). *Match*

relatively stable pattern of vegetation over the c. 70-year time span. Most of the changes are in the right foreground where broadleaved species (probably cottonwoods) have disappeared and ponderosa pine is now more prominent. The gray tones on the slope to the right are crowns of Douglas fir defoliated by spruce budworm. Willow shrubs have established the far shoreline.

Plate 65. Horseshoe Park and Fall River Valley. *Original*

Location: This west-northwest view of Mt. Chapin was taken 0.8 km southeast of the intersection of sections 11 and 13, T4N R74W, at an elevation of 2593 m.

Original: 1916, W. T. Lee No. 992, USGS.
Match: 1986, T. T. Veblen and D. C. Lorenz No. R4.

Plate 65. Horseshoe Park and Fall River Valley. *Match*

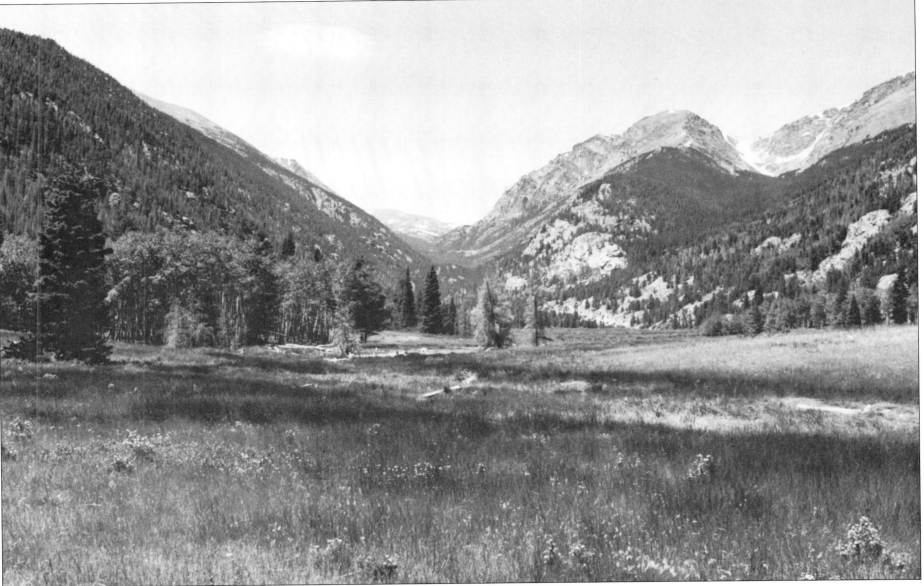

Description: Other than the removal of the cabin and fence and the cutting of the aspen, these photographs show relatively little change over the c. 70-year time span. There has been some increase in forest density on the northerly slope in the left center.

Plate 66. South Lateral Moraine and Longs Peak. *Original*

Location: This is a view south of South Lateral Moraine from Moraine Park. The camera point was 0.3 km southwest of the intersection of sections 28 and 32, T5N R73W, at an altitude of 2501 m.
Original: c. 1885, RMNP.

Match: 1986, T. T. Veblen and D. C. Lorenz No. R2.
Description: Moraine Park was the location of

Plate 66. South Lateral Moraine and Longs Peak. *Match*

several hotels and numerous homes prior to the creation of Rocky Mountain National Park. The original photograph shows that by the 1880s the forest on the moraine had been extensively burned and logged. The modern photograph shows recovery to a dense forest of mainly lodgepole pine. Soon after the creation of Rocky Mountain National Park, the buildings in the foreground were dismantled.

Plate 67. Steads's Ranch in Moraine Park. *Original*

Location: This north view of Beaver Mountain was taken 0.6 km southwest of the intersection of sections 28 and 32, T5N R73W, at an altitude of 2501 m.

Original: c. 1920, L. C. McClure No. 2626, DPL.
Match: 1986, T. T. Veblen and D. C. Lorenz No. R15.

Description: The background slopes in the original photograph are mostly covered with open stands of ponderosa pine and Douglas fir and small clumps of aspen. At the right center is a dense

Plate 67. Steads's Ranch in Moraine Park. *Match*

stand of lodgepole pine on the north-facing slope. The match indicates considerable increase in the density of the ponderosa pine and Douglas fir stands and a substantial decline in the abundance of aspen. The gray tones in the center are crowns of Douglas fir defoliated by spruce budworm. All vestiges of Steads's ranch have been removed by the Park Service.

Plate 68. Lake Irene, Milner Pass. *Original*

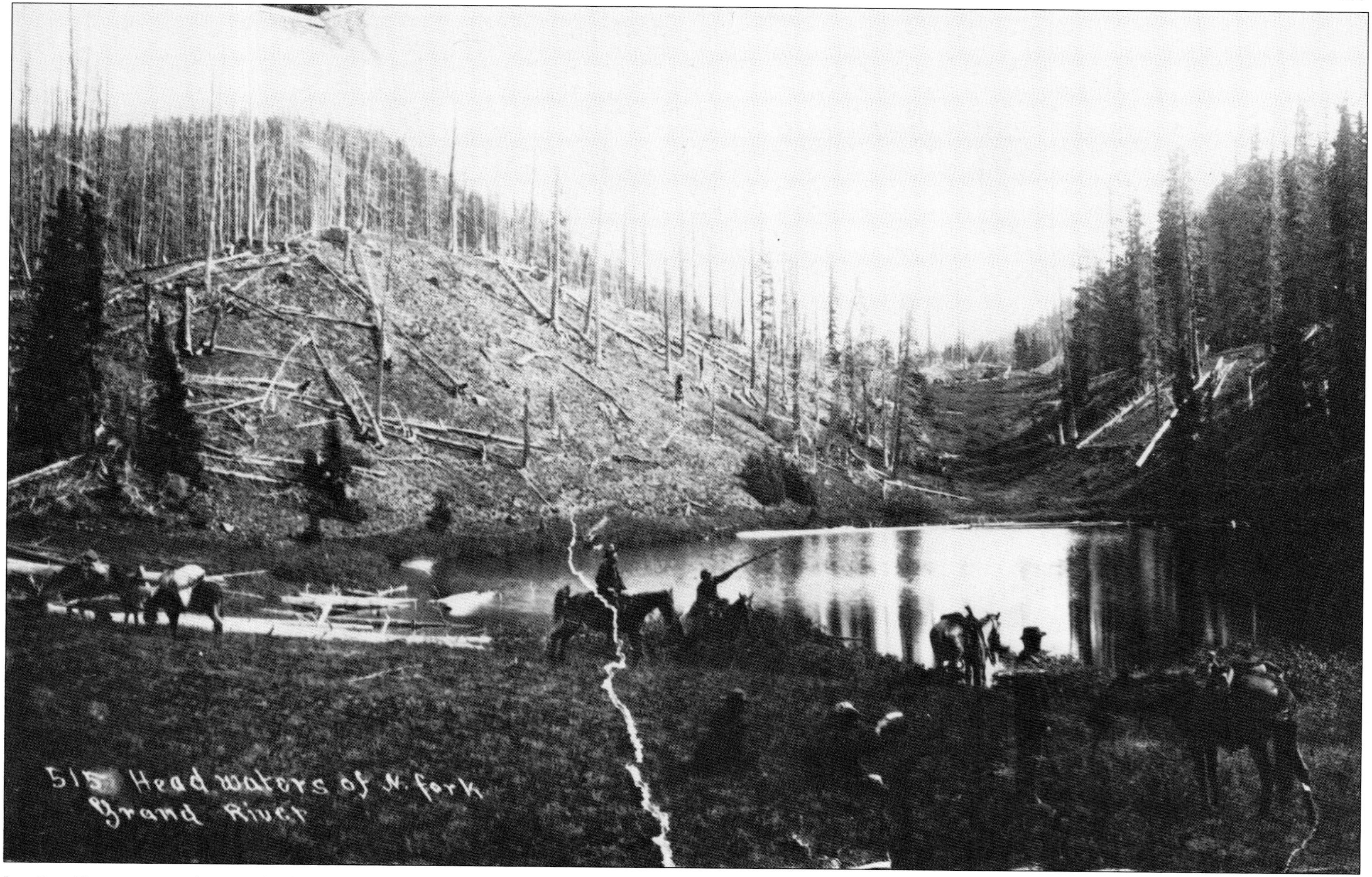

Location: The camera point was 0.6 km southeast of the intersection of sections 6 and 7, T5N R75W, at an elevation of 3233 m.

Original: c. 1890, unknown photographer, No. 515, RMNP.

Match: 1986, T. T. Veblen and D. C. Lorenz No. R22.

Plate 68. Lake Irene, Milner Pass. *Match*

Description: The original photograph shows that the forests on the northern side of Lake Irene had been burned by the 1890s. The match shows that recovery on the southerly slope has been partial. This is a rare example of an originally forested site that is today a grassland. The high elevation and north-facing slope may explain the minimal tree regeneration.

Plate 69. Lulu City and Lulu Mountain. *Original*

Location: This view north of Lulu Mountain was taken 0.6 km southwest of the intersection of sections 30 and 31, T6N R75W, at an elevation of 2882 m.

Original: 1905, unknown photographer, No. 05315, DPL.
Match: 1986, T. T. Veblen and D. C. Lorenz No. R23.

Description: Lulu City, a gold and silver mining town, was founded in 1880 and boomed for only a few years before nearly disappearing in the 1890s. The original photograph shows that most of the

Plate 69. Lulu City and Lulu Mountain. *Match*

slopes surrounding the town had been burned by the 1880s. The match shows recovery to dense stands dominated mainly by lodgepole pine mixed with Engelmann spruce and subalpine fir. The townsite, which once supported a population of 200, is now grassland with scattered Engelmann spruce, subalpine fir, and lodgepole pine. The bare area on the upper slope in the background is a recently constructed aqueduct, the Grand Ditch.

Discussion and Conclusions

Several patterns of vegetation change in the Front Range are evident in the matched photographs. In some cases the ecological processes responsible for the change are well understood but in many instances the explanations of the changes are complex and controversial.

A major pattern evident in the historical photographs is the widespread importance of fire prior to c. 1920. Most of the montane zone and a large part of the subalpine zone had been burned during the latter part of the nineteenth century. Thus, stands of dead standing, charred trees are commonly depicted in the historical photographs. Most of these fires appear to have been set, intentionally or accidentally, by white settlers during the period of intensive prospecting beginning about 1860 and with the initiation of livestock raising. Earlier, of course, there had been frequent fires ignited by lightning or the Native American population, but fire-history studies show a substantial increase in fire frequency during the settlement period (Rowdabaugh 1978; Laven et al. 1980). Fires were particularly frequent and extensive in the montane zone of Boulder County where today we find primarily young postfire stands of ponderosa pine and Douglas fir (Veblen and Lorenz 1986).

Although also affected by increased fire frequency during the settlement period, the montane forests of Rocky Mountain National Park are not as uniformly young as those of Boulder County (Peet 1981). This reflects the fact that Boulder County is located in the mineralized belt of the Front Range while mining in Larimer County (much of which is in Rocky Mountain National Park) was relatively minor. Similarly, the degree of logging for fuel, mine props, and town and railroad construction in Boulder County was substantially greater than it was in the present area of Rocky Mountain National Park. Nevertheless, a surprisingly large percentage of the forests of Rocky Mountain National Park suffered at least light logging as reflected by the abundance of cut stumps over large parts of the Park.

In the subalpine zone, logging and burning during the settlement period were also widespread even though these disturbances were not as extensive as they were in the montane zone. In Rocky Mountain National Park, the extent of burned forest at the turn of the century in some subalpine areas such as Bear and Bierstadt Lakes is startling (e.g., Plates 57–62). Nevertheless, the presence of many stands of Engelmann spruce and subalpine fir dominated by numerous trees at least 200 years old is evidence that much of the Front Range subalpine zone escaped stand-devastating fires during the settlement period (Peet 1981; Veblen 1986a; Veblen et al. 1989).

At the time most of the historical photographs were taken (about 1880–1915), the forests of the montane zone of the Front Range were in early stages of stand development due to the extensive logging and burning of the Euro-American settlement period. Today, these stands contain even-aged tree populations in which most trees are 60 to 100 years old (Fig. 12). In some cases in the

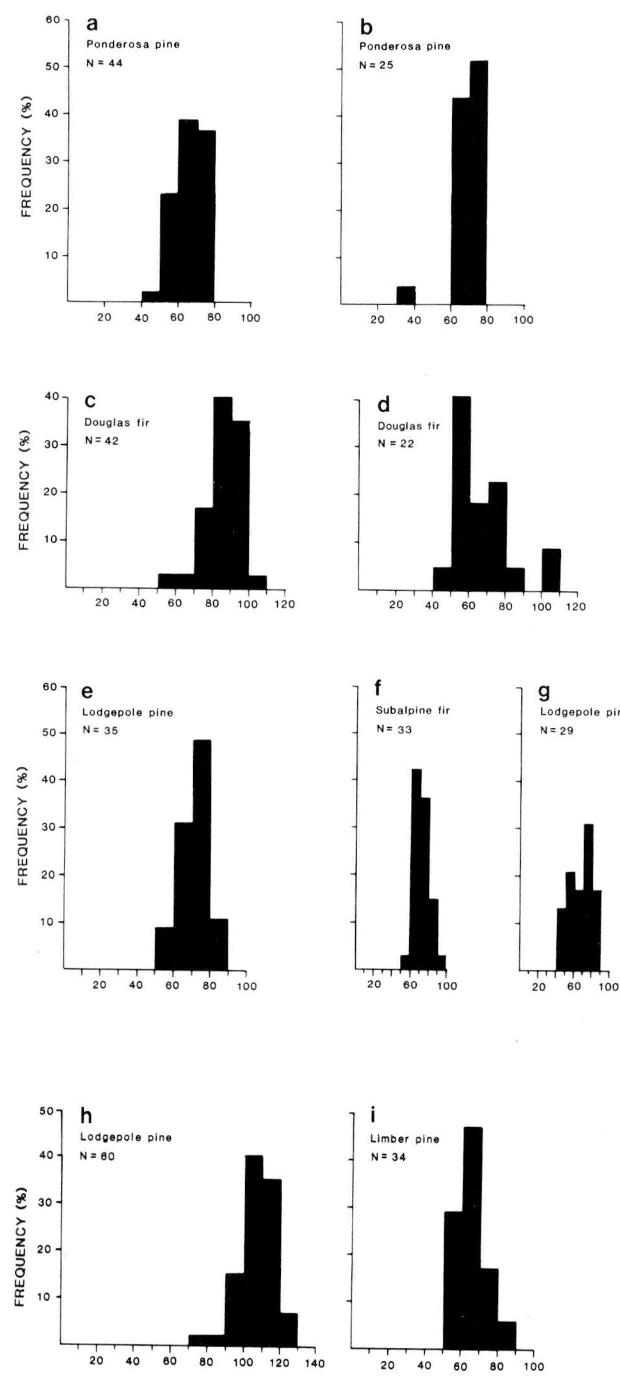

Figure 12. Typical examples of tree-age frequency distributions in montane forests in Boulder County. A former grassland site which has been invaded by ponderosa pine is illustrated by Figure 12a. All other tree populations are typical of postfire stands which initiated following fires in the mid- to late-nineteenth century. From Veblen and Lorenz (1986).

montane zone, these early stages of stand development were dominated by species other than the dominants of the burned forests, but in most cases the original dominants have regenerated following the fire (Veblen and Lorenz 1986). For example, on xeric sites comparison of dead and remnant trees with the composition of postfire populations shows some tendency for a shift in dominance from Douglas fir to ponderosa pine and/or limber pine following fire (e.g., Plates 21 and 44). At more mesic sites, stands originally dominated by Douglas fir have regenerated to the same species after fire (e.g., Plates 13 and 42). In contrast, in the lower part of the subalpine zone where today lodgepole pine is dominant, fire has had a greater impact on the species composition of the present forests. Burning of Engelmann spruce and subalpine fir forests on relatively dry sites has resulted in extensive areas of lodgepole pine–dominated forests (e.g., Plates 26 and 69). Aspen is also common in some of the young postfire stands, and the matched photographs show a tendency for this species to be replaced by conifers during the subsequent relatively fire-free period (e.g., Plates 27 and 41). In other cases, however, dead and remnant trees indicate that the prefire forest was also dominated by lodgepole pine. Thus, the presettlement forest was not uniformly in an old-growth stage dominated by shade-tolerant species. Although the evidence of forest age structure and repeat photography indicates both successional replacement and regeneration of the original dominants following fire for different sites, there is insufficient quantitative data on Front Range forests to be sure which pattern prevailed.

The rates of forest recovery following fire vary in a predictable way in relation to elevation and moisture status of the habitat. The most rapid rates of recovery occur on moist sites near the center of the elevational distribution of a particular forest type (Peet 1981; Veblen and Lorenz 1986). Conversely, towards either the upper or lower elevational limit of a forest type or on a more xeric site, the time required for recovery, as measured either by return to the same species composition or to the same level of basal area, is slower. On dry south-facing slopes near timberline, recovery following fire can be exceedingly slow. Here, where stands originally dominated by

Engelmann spruce and subalpine fir are burned, the site may be dominated by limber or lodgepole pine for more than 200 years before eventually returning to dominance by spruce and fir (Peet 1981; Veblen 1986a).

Following the massive burning of montane forests during the Euro-American settlement period, fire frequency declined dramatically. Much of this decline is due to modern fire suppression efforts, but it is also partially due to the change in forest structure resulting from the nineteenth-century fires. Young stands of lodgepole pine in the upper montane and lower subalpine zones are resistant to fire due to the lack of accumulation of fuels (Romme 1982). Thus, the increase of young lodgepole pine stands due to nineteenth-century burning has created more fire-resistant stands for this forest type. An important consequence of decreased fire frequency in the montane zone is the increase in stand density in ponderosa pine and Douglas fir stands. Previously, more frequent surface fires would destroy many seedlings and saplings of these species. During the past half century of fire control these have survived to become suppressed, slow-growing, small trees. Thus, in stands long free of fire, small trees occur in much greater densities than what would have been expected during presettlement times when surface fires periodically "thinned" these stands.

The presence of dense populations of stagnant Douglas fir beneath the main canopies in montane stands dramatically changes the susceptibility of these forests to disturbance by either fire or insect. The lack of frequent surface fires permits an accumulation of fuel and the abundance of small trees that serve as "fire ladders." The result is an increased likelihood that a fire once ignited will become a stand-devastating crown fire. Although fire control has undoubtedly decreased the frequency of fires in the montane zone, the resulting changes in stand structure have increased the probable severity of any given fire. Given the abundance of residences in the montane zone of the Front Range, this change in the fire regime is a particularly serious management concern.

The abundance of stagnant, small Douglas fir in the understories of montane forests may have also increased the severity of spruce budworm outbreaks. Small, slow-growing trees are more likely to be killed by spruce budworm. The greater amount of food (Douglas fir needles) presented by denser stands may also result in larger populations of budworm, which in turn may cause more damage to the large canopy trees. Although spruce budworm epidemics also affected these forests long before white settlement and fire suppression, there is some tree-ring evidence that epidemics over the past few decades have been more extensive and more severe (Swetnam 1987). Much further research is required before definitive conclusions can be reached, but it appears that fire suppression and the resulting increase in abundance of suppressed, slow-growing Douglas fir have exacerbated the spruce budworm problem in the Front Range.

The effects of fire suppression on bark beetle outbreaks, such as the mountain pine beetle, are less clear. Decreased fire frequency has undoubtedly created montane forests in which competition is substantially greater due to increased stand density. Thus, there are more slow-growing trees. A tree's ability to resist bark beetle attack is largely determined by its vigor (Larrson et al. 1983). So, fast-growing trees not suffering from intense competition are the most likely to survive a bark beetle attack by "pitching out" the beetles. Conversely, weak, slow-growing trees are more likely to be successfully attacked and killed by bark beetles. On the other hand, for a successful attack bark beetles require large volumes of cambium for food and shelter, which makes larger-diameter trees more likely victims. So, by creating stands of relatively small diameters, the extensive nineteenth-century fires may have decreased the proportion of trees susceptible to mountain pine beetle attacks. Decreased fire frequency during the present century probably has increased the severity of mistletoe infestation in montane forests. Weakening of trees by mistletoe, in turn, makes them more susceptible to attack by insects such as bark beetles. However, virtually nothing is known about the frequency or severity of presettlement mountain pine beetle epidemics, and, consequently, it is difficult to say whether they have become more of a problem in the present century.

In the case of spruce beetle, we know that the most extensive outbreak in western

Colorado occurred in the mid-nineteenth century long before fire suppression altered the natural fire regime (Baker and Veblen 1990). The same area was also affected by a devastating outbreak in the 1940s (Schmid and Frye 1977). Given the natural fire interval of at least 200 years in Engelmann spruce and subalpine fir forests (Romme and Knight 1981), fire suppression is not the likely explanation for the most recent outbreak. Matched photographs of subalpine forests in the Front Range show similar numbers of dead standing trees 80 years ago as compared to the present forests. Although the cause of death in the original photographs is not known, some of the mortality was probably due to spruce beetles. The photographs do not imply that Front Range subalpine old-growth forests today are characterized by more tree mortality than they were at the end of the last century.

Another major pattern revealed by the matched photographs is the increased density of conifers, particularly in the montane zone. Stands that were initially open woodlands are now dense forests, and where the vegetation was originally grassland we often find young populations of ponderosa pine (e.g., Plates 6 and 19 and Fig. 12b). In many cases, particularly in the lower foothills, the expansion and increase in density of ponderosa pine has been dramatic. Similar conifer invasions of grasslands, usually initiated between the 1870s and the first decades of the present century, are widespread throughout the western United States (Fonda and Bliss 1969; Vankat and Major 1978; Vale 1981; Agee and Dunwiddie 1984; Savage 1989). There are a variety of possible explanations for these changes. One possible explanation is a change towards a more mesic climate due to changes in precipitation and/or temperature, so that the competitive balance is shifted in favor of ponderosa pine over more drought-tolerant grasses. For ponderosa pine in Arizona there are clear correlations between periods of abundant tree establishment and times of greater moisture availability (Pearson 1934; Cooper 1960). Overgrazing by livestock may also favor ponderosa pine invasion of grasslands by reducing the competition from grass and often exposing bare mineral soil for tree seedling establishment (Rummell 1951). For the Front Range, overgrazing by cattle in the late nineteenth century has been suggested as the triggering mechanism for the conifer invasion (Marr 1961). However, in recent decades despite light or no grazing at many sites in the Front Range, the invasion by ponderosa pine has continued (Veblen and Lorenz 1986). A third explanation is simply the decreased fire frequency of the present century. Frequent surface fires during the pre-Euro-American settlement period probably prevented ponderosa pine from occurring at many sites where it is currently abundant. Particularly, where open woodlands of ponderosa pine have become dense forests, the reduction in frequency of surface ground fires appears to be the most plausible explanation for the observed change. However, it is likely that all three factors—climatic variability, overgrazing, and decreased fire frequency—have played some role in the invasion of Front Range grasslands by ponderosa pine.

Although repeat photography leaves many unanswered questions about the ecological causes of vegetation change, it clearly documents the often dramatic vegetation changes that have occurred over the past hundred years in the Colorado Front Range. These changes are clearly related to the equally dramatic social and economic changes that have occurred in the Front Range over the past century. Until half a century ago, humans were massively altering the Front Range forests by logging, burning, and the raising of livestock. While these activities continue on a small scale today, their intensities are only small fractions of what they were in the nineteenth-century period of mining and pioneer ranching. In recent decades, the forests of the Front Range have become much less important as a source of timber and instead are used primarily for recreational and residential purposes. While these uses do not result in rapid forest changes comparable to those of the nineteenth century, these less exploitative uses, because of the associated policy of fire suppression, may also result in major long-term changes in the character of the forests. Changes in forest structure due to fire suppression over the last 60 years or so are already evident and in some forest types appear to have contributed to more severe outbreaks of forest insect pests. The historical perspective provided by repeat photography allows us to better appreciate the constantly changing nature of the Front Range forests.

References Cited

Agee, J. K. and P. W. Dunwiddie. 1984. Recent forest development on Yellow Island, San Juan County, WA. *Canadian Journal of Botany* 62:2074–80.

Alexander, R. R. 1974. *Silviculture of subalpine forests in the southern Rocky Mountains: The status of our knowledge.* U.S. Forest Service Research Paper RM-121. 88 p.

Alexander, R. R. 1986. *Silvicultural systems and cutting methods for ponderosa pine forests in the Front Range of the central Rocky Mountains.* U.S. Forest Service General Technical Report RM-128. 22p.

______, R. C. Shearer, and W. D. Shepperd. 1984. *Silvical characteristics of subalpine fir.* U.S. Forest Service General Technical Report RM-115. 29 p.

______, and W. D. Shepperd. 1984. *Silvical characteristics of Engelmann spruce.* U.S. Forest Service General Technical Report RM-114. 38 p.

Baker, W. L. and T. T. Veblen. 1990. Spruce beetles and fires in the nineteenth century subalpine forests of western Colorado. *Arctic and Alpine Research* 22:65–80.

Barbour, M. G., J. H. Burk, and W. D. Pitts. 1987. *Terrestrial plant ecology.* The Benjamin/Cummings Publishing Company, Menlo Park. 634 p.

Barrows, J. S. 1936. Forest insect problems of the Rocky Mountain National Park. U.S. Department of the Interior, National Park Service, unpublished report. 15 p.

Barry, R. G. 1972. *Climatic environment of the east slope of the Colorado Front Range.* Institute of Arctic and Alpine Research, University of Colorado, Occasional Paper 3. 206 p.

Benedict, J. B. 1973. Chronology of cirque glaciation, Colorado Front Range. *Quaternary Research* 3:585–99.

______. 1974. Early occupation of the Caribou Lake site, Colorado Front Range. *Plains Anthropologist* 63:1–4.

______. 1975. The Murray site: A late prehistoric game drive system in the Colorado Rocky Mountains. *Plains Anthropologist* 69:161–74.

Buccholtz, C. W. 1983. *Rocky Mountain National Park: A history.* Colorado Associated University Press, Boulder. 255 p.

Campbell, R. W. 1986. Population dynamics. *In* M. H. Brookes, R. W. Campbell, J. J. Colbert, R. G. Mitchell, and R. W. Stark, eds., *Western spruce budworm,* p. 71–88. U.S. Forest Service Technical Bulletin No. 1694. 198 p.

Carlson, C. E., D. G. Fellin, and W. C. Schmidt. 1982. The western spruce budworm in northern Rocky Mountain forests: A review of ecology, insecticidal treatments and silvicultural practices. *In* J. O'Loughlin and R. D. Pfister, eds., *Management of second growth forests: The state of knowledge and research needs,* p. 76–103. University of Montana, Missoula. 201 p.

Cates, R. G., and H. Alexander. 1982. Host resistance and susceptibility. *In* J. B. Mitton and K. B. Sturgeon, eds., *Bark beetles in North American conifers,* p. 212–63. University of Texas Press, Austin. 527 p.

Chase, A. 1986. *Playing God in Yellowstone: The destruction of America's first national park.* Atlantic Monthly Press, Boston. 304 p.

Chronic, J., and H. Chronic. 1972. *Prairie, peak, and plateau: A guide to the geology of Colorado.* Colorado Geological Survey Bulletin 32. 126 p.

Clements, F. E. 1910. The life history of lodgepole burn forests. *U.S. Forest Service Bulletin* 79:7–56.

______. 1916. *Plant succession: An analysis of the development of vegetation.* Carnegie Institute of Washington, Washington, D.C., Publication 242. 512 p.

Cole, W. E. and G. D. Amman. 1969. *Mountain pine beetle infestations in relation to lodgepole pine diameters.* U.S. Forest Service Research Note INT-95. 7 p.

Cooper, C. F. 1960. Changes in vegetation, structure, and growth of southwestern pine forests since white settlement. *Ecological Monographs* 30:129–64.

Crane, M. F. 1982. Fire ecology of Rocky Mountain region forest habitat types. U.S. Forest Service, Region Two, unpublished report. 266 p.

Crossen, F. 1978. *The Switzerland Trail of America.* Pruett Press, Boulder. 422 p.

Drury, W. H., and I. C. T. Nisbet. Inter-relations between developmental models in geomorphology, plant ecology, and animal ecology. *General Systems* 16:57–68.

Fonda, R. W., and L. C. Bliss. 1969. Forest vegetation of the montane and subalpine zones, Olympic

Mountains, Washington. Ecological Monographs 39:271–301.

Fossett, F. 1880. *Colorado—Its gold and silver mines, farms and stock ranges, and health and pleasure resorts.* C. G. Crawford, New York. 624 p.

Fritz, P. S. 1933. Mining districts of Boulder County, Colorado. Ph.D. thesis, University of Colorado, Boulder. 639 p.

Furniss, R. L., and V. M. Carolin. 1977. *Western forest insects.* U.S. Forest Service Miscellaneous Publication 1339. 654 p.

Glidden, D. E. 1982. *Winter wind studies in Rocky Mountain National Park.* Rocky Mountain Nature Association, Denver. 29 p.

Griffiths, M., and L. Rubright. 1983. *Colorado.* Westview Press, Boulder. 325 p.

Gruell, G. E. 1980. *Fire's influence on wildlife habitat on the Bridger-Teton National Forest, Wyoming.* Volume 1—*Photographic record and analysis.* U.S. Forest Service Research Paper INT-235. 207 p.

Hafen, L. R., and A. W. Hafen. 1956. *Rufus B. Sage, his letters and papers, 1836–1847.* Volume 2. The Arthur H. Clark Company, Glendale. 312 p.

Hansen, W. R., J. Chronic, and J. Matelock. 1978. *Climatography of the Front Range urban corridor and vicinity, Colorado.* U.S. Geological Survey Professional Paper 1019. 59 p.

Harrison, A. E. 1974. Reoccupying unmarked camera stations for geological observations. *Geology* 2:469–71.

Hastings, J. R., and R. M. Turner. 1965. *The changing mile.* University of Arizona Press, Tucson. 317 p.

Hinds, T. E. 1976. *Aspen mortality in Rocky Mountain campgrounds.* U.S. Forest Service Research Paper RM-164. 20 p.

———, and F. G. Hawksworth. 1966. *Indicators and associated decay of Engelmann spruce in Colorado.* U.S. Forest Service Research Paper RM-25. 15 p.

Hughes, J. D. 1977. *American Indians of Colorado.* Pruett, Boulder. 145 p.

Husted, W. 1965. Early occupation of the Front Range. *American Antiquity* 30:494–98.

Ives, J. D. 1983. Introduction: A description of the Front Range. In *Geoecology of the Colorado Front Range: A study of alpine and subalpine environments,* p. 1–8. Westview Press, Boulder. 484 p.

Johnson, D. D., and A. J. Cline. Colorado mountain soils. *Advances in Agronomy* 17:233–81.

Jones, W. C., and E. B. Jones. 1983. *Photo by McClure: The railroad, cityscape and landscape photographs of L. C. McClure.* Pruett, Boulder. 254 p.

Keen, F. P. 1940. Longevity of ponderosa pine. *Journal of Forestry* 38:597–98.

Kemp, D. C. 1960. *Silver, gold and black iron: A story of the Grand Island Mining District of Boulder County, Colorado.* Sage Swallow, Denver. 230 p.

Knapp, A. K. and W. K. Smith. 1981. Water relations and succession in subalpine conifers in southeastern Wyoming. *Botanical Gazette* 142:502–11.

———. 1982. Factors influencing understory seedling establishment of Engelmann spruce (*Picea engelmannii*) and subalpine fir (*Abies lasiocarpa*) in southeastern Wyoming. *Canadian Journal of Botany* 60:2753–61.

Larrson, S., R. Oren, R. H. Waring, and J. W. Barrett. 1983. Attacks of mountain pine beetle as related to tree vigor of ponderosa pine. *Forest Science* 29:395–402.

Laven, R. D., P. N. Omi, J. G. Wyant, and A. S. Pinkerton. 1980. Interpretation of fire scar data from a ponderosa pine ecosystem in the central Rocky Mountains, Colorado. *In* M. A. Stokes and J. H. Dieterich, eds., *Proceedings of the fire history workshop, October 20–24, 1980, Tucson, Arizona,* p. 46–49. U.S. Forest Service General Technical Report RM-81. 142 p.

Lotan, J. E. 1975. The role of cone serotiny in lodgepole pine forests. *In* D. M. Baumgartner, ed., *Management of lodgepole pine ecosystems,* p. 471–95. Washington State University, Pullman. 519 p.

———, B. M. Kilgore, W. C. Fischer, and R. W. Mutch, eds. 1985. *Proceedings: Symposium and workshop on wilderness fire, Missoula, Montana, November 15–18, 1983.* U.S. Forest Service General Technical Report INT-182. 434 p.

Lovering, T. S., and E. N. Goddard. 1950. *Geology and ore deposits of the Front Range, Colorado.* U.S. Geological Survey Professional Paper 223. 319 p.

Malde, H. E. 1973. Geologic bench marks by terrestrial photography. *U.S. Geological Survey Journal of Research* 1:193–206.

Marr, J. W. 1961. *Ecosystems of the east slope of the Front Range in Colorado.* University of Colorado Studies Series in Biology 8. 134 p.

———. 1964. The vegetation of the Boulder area. *In* H. G. Rodeck, ed., *Natural history of the Boulder area,* p. 34–42. University of Colorado Museum Leaflet 13. 178 p.

McCambridge, W. F., and G. C. Trostle. 1972. *The mountain pine beetle.* U.S. Forest Service, Forest Pest Leaflet 2. 6 p.

Miles, J. 1979. *Vegetation dynamics.* Chapman and Hall, London. 80 p.

Muir, P. S., and J. E. Lotan. 1985. Disturbance history and serotiny of *Pinus contorta* in western Montana. *Ecology* 66:1658–68.

Parker, A. J., and K. C. Parker. 1983. Comparative successional roles of trembling aspen and lodgepole pine in the southern Rocky Mountains. *Great Basin Naturalist* 43:447–55.

Pearson, G. A. 1934. Grass, pine seedlings, and grazing. *Journal of Forestry* 32:545–55.

Peet, R. K. 1978. Latitudinal variation in southern Rocky Mountain forests. *Journal of Biogeography* 5:275–89.

———. 1981. Forest vegetation of the Colorado Front Range: Composition and dynamics. *Vegetation* 45:3–75.

———. 1988. Forests of the Rocky Mountains. *In* M. G. Barbour and W. D. Billings, eds., *North American terrestrial vegetation,* p. 63–102. Cambridge University Press, New York. 434 p.

Pettem, S. 1980. *Red Rocks to riches.* Stonehenge, Boulder. 123 p.

Pickett, S. T. A., and P. S. White, eds. 1985. *The ecology of natural disturbance and patch dynamics.* Academic Press, Orlando. 472 p.

Richmond, G. M. 1960. Glaciation of the east slope of Rocky Mountain National Park, Colorado. *Geological Society of America Bulletin* 71:1371–82.

Roe, A. L., and G. D. Amman. 1970. *The mountain pine beetle in lodgepole pine forests.* U.S. Forest Service Research Paper INT-71. 23 p.

Rogers, G. 1982. *Then and now: A photographic history of vegetation change in the central Great Basin Desert.* University of Utah Press, Salt Lake City. 188 p.

______, H. Malde, and R. Turner. 1984. *Bibliography of repeat photography for evaluating landscape change.* University of Utah Press, Salt Lake City. 179 p.

Romme, W. H. 1982. Fire and landscape diversity in subalpine forests of Yellowstone National Park. *Ecological Monographs* 52:199–221.

______, and D. H. Knight. 1981. Fire frequency and subalpine forest succession along a topographic gradient in Wyoming. *Ecology* 62:319–26.

Rowdabaugh, K. M. 1978. The role of fire in the ponderosa pine-mixed conifer ecosystems. Master's thesis. Colorado State University, Fort Collins. 121 p.

Rummell, R. 1951. Some effects of livestock grazing on ponderosa pine forest and range in central Washington. *Ecology* 32:594–607.

Ruxton, G. F. 1951. *Life in the Far West.* University of Oklahoma Press, Norman. 252 p.

Savage, M. 1989. Structural dynamics of a pine forest in the American southwest under chronic human disturbance. Ph.D. thesis, University of Colorado, Boulder. 198 p.

Schmid, J. M., and R. C. Beckwith. 1972. *The spruce beetle.* U.S. Forest Service Forest Pest Leaflet 127.

Schmid, J. M., and R. H. Frye. 1977. *Spruce beetle in the Rockies.* U.S. Forest Service General Technical Report RM-49. 38 p.

Schoolland, J. B. 1982. *Boulder: Then and now.* Johnson, Boulder. 285 p.

Schubert, G. H. 1974. *Silviculture of southwestern ponderosa pine: The state of our knowledge.* U.S. Forest Service Research Paper RM-123. 39 p.

Shea, K. L., and M. C. Grant. 1986. Clonal growth in spire-shaped Engelmann spruce and subalpine fir trees. *Canadian Journal of Botany* 64:255–305.

Sheppard, R. F. 1959. Phytosociological and environmental characteristics of outbreak and non-outbreak areas of the two year cycle spruce budworm, *Choristoneura fumiferana*. *Ecology* 40:608–20.

Skinner, T. V., and R. D. Laven. 1983. A fire history of the Longs Peak region of Rocky Mountain National Park. In *Proceedings of the seventh conference on fire and forest Meteorology, Fort Collins,* p. 71–74.

Smith, P. 1981. *A look at Boulder from settlement to city.* Pruett, Boulder. 263 p.

Stevens, R. E., J. W. Brewer, and D. A. Leatherman. 1980. *Insects associated with ponderosa pine in Colorado.* U.S. Forest Service General Technical Report RM-75. 39 p.

Stewart, O. C. 1956. Fire as the first great force employed by man. *In* W. L. Thomas, ed., *Man's role in changing the face of the earth,* p. 115–33. University of Chicago Press, Chicago. 1193 p.

Swetnam, T. W. 1987. A dendrochronological assessment of western spruce budworm, *Choristoneura occidentalis* Freeman, in the southern Rocky Mountains. Ph.D. thesis. University of Arizona, Tucson. 213 p.

Tansley, A. G. 1935. The use and abuse of vegetational concepts and terms. *Ecology* 16:284–307.

Tice, J. H. 1872. *Over the plains and on the mountains.* Industrial Age Printing, St. Louis. 262 p.

Vale, T. R. 1981. Tree invasion of montane meadows in Oregon. *American Midland Naturalist* 105:61–69.

______. 1987. Vegetation change and park purposes in the high elevations of Yosemite National Park, California. *Annals of the Association of American Geographers* 77:1–18.

Vankat, J. L., and J. Major. 1978. Vegetation changes in Sequoia National Park, California. *Journal of Biogeography* 5:377–402.

Veblen, T. T. 1986a. Age and size structure of subalpine forests in the Colorado Front Range. *Bulletin of the Torrey Botanical Club* 113:225–40.

______. 1986b. Treefalls and the coexistence of conifers in subalpine forests of the central Rockies. *Ecology* 67:644–49.

______. In press. Regeneration dynamics. *In* D. C. Glenn-Lewin, R. K. Peet, and T. T. Veblen, eds., *Plant succession.* Chapman Hall, London.

______, and D. C. Lorenz. 1986. Anthropogenic disturbance and recovery patterns in montane forests, Colorado Front Range. *Physical Geography* 7:1–24.

______, and D. C. Lorenz. 1988. Recent vegetation changes along the forest/steppe ecotone in northern Patagonia. *Annals of the Association of American Geographers* 78:93–111.

______, K. S. Hadley, M. S. Reid, and A. J. Rebertus. 1989. Blowdown and stand development in a Colorado subalpine forest. *Canadian Journal of Forest Research* 19:1218–25.

______, F. M. Schlegel, and B. Escobar. 1980. Structure and dynamics of old-growth *Nothofagus* forests in the Valdivian Andes, Chile. *Journal of Ecology* 68:1–31.

Weber, W. A. 1976. *Rocky Mountain flora.* Colorado Associated University Press, Boulder. 479 p.

Whipple, S. A., and R. L. Dix. 1979. Age structure and successional dynamics of a Colorado subalpine forest. *American Midland Naturalist* 101:142–58.

White, P. S. 1979. Pattern, process, and natural disturbance in vegetation. *Botanical Review* 45:229–99.

Whittaker, R. H. 1975. *Communities and ecosystems.* MacMillan, New York. 385 p.

Wolle, M. S. 1949. *Stampede to timberline.* Sage, Denver. 544 p.

Index